THE HUMAN EXPERIMENT

TWO YEARS AND TWENTY MINUTES INSIDE

BIOSPHERE 2

JANE   POYNTER

THUNDER'S MOUTH PRESS

NEW YORK

THE HUMAN EXPERIMENT:

Two Years and Twenty Minutes Inside Biosphere 2

Published by

Thunder's Mouth Press

An Imprint of Avalon Publishing Group Inc.

245 West 17th Street, 11th Floor

New York, NY 10011

AVALON
publishing group incorporated

First Printing, September 2006

Library of Congress Cataloging-in-Publication Data is available.

ISBN-10: 1-56025-775-X
ISBN-13: 978-1-56025-775-2

Book design by Maria E. Torres

Printed in the United States of America

Distributed by Publishers Group West

To all whose passion realized the dream
of Biosphere—you are heroes.
And to Roy Walford, my inspiration, my friend.

CONTENTS

PROLOGUE

Eight o' clock in the morning on September 26, 1993. I stood in my prickly blue jumpsuit with the other seven inmates of the Bubble, as some of us liked to call it. We waited for the radio announcement that it was time to walk through the double-doored airlock, that the mission was finally over. I would like to say that I was pondering heady thoughts about the future of mankind, but all I could think of was how much I wished that dear Jane Goodall would shut up.

I have the deepest respect for my fellow countrywoman who has dedicated her life to the study and conservation of chimpanzees, taught us that apes use tools and laugh, too, and caused us to redefine what it is that makes us human. But as Jane gave the keynote speech leading to our reentry into the world, into what we called Biosphere 1, the minutes ticked by with agonizing slowness.

"Come on, people," I muttered to myself. "I signed up for two years—not two years and one minute, or two minutes. Only two years."

Eight ten. "Jane, let us apes out of the cage!"

Eight fifteen.

Finally, some screeching over the radio told us to scurry to the heavy metal airlock fashioned out of submarine bulkheads years earlier. It was eight twenty. We stepped in, the door swinging closed behind us with the bang and scrape of the closure mechanism, and the outer door opened.

One by one, we stepped out of our simple life of milking goats and weeding the garden, of weather reports that included—along with temperature and humidity readings—the levels of oxygen and carbon dioxide. We left the daily struggle with tedium and discord, and stepped into applause, a trumpeted fanfare, cheering, a sea of cameras, backslaps, and handshakes.

For the first time in two years and twenty minutes, I inhaled the view of the bright desert sky with no white bars dissecting it into geometric patterns. The air seemed thin, insipid, not the thick atmosphere redolent with molecules from plants, fungi, animals—the pungent, pleasant, and unmistakable, earthy fragrance of Biosphere 2.

Since September 26, 1991, the eight of us had risked much to live as if on Mars, farming all our food, recycling our water, our waste, and even the oxygen we breathed in our hermetically sealed 3.15-acre world. The rainforest, savannah, desert, ocean, and marsh had been our in-vitro test subjects for ecological research. But the glass and steel structure made a pressure cooker, our human foibles boiling to the surface in what some named the Human Experiment.

Now ten years have smoothed the searing anger I felt upon completing our mission. I can at last recall and assess the controversy and all that occurred in and around our New Age Garden of Eden, aided by the many people I have interviewed and the records I have read, with what approximates objectivity. I want to tell this extraordinary adventure tale of a colorful band of mavericks attempting what many said was impossible to set the record straight, to halt the maelstrom of misinformation that still swirls around the project.

Even now, when I open boxes containing books and clothing I had with me during my sojourn, the smell emanating from the brown cardboard transports me instantaneously back inside. . . .

Chapter 1

OXYGEN DILEMMA

I t was early April 1992, and I sat at the small oak desk in my bright, comfortable living quarters in Biosphere 2. Mine was like the other seven biospherians'—split-level with a small mezzanine where my queen-size bed stood. Down a spiral staircase lay my roughly fifteen-by-fifteen-foot living room where I sat writing in my journal, as I did every evening I could summon enough energy. The room was my refuge.

Behind me sat a teal-colored sofa and chair, and a small oak entertainment center that held a television and tape player, which was filling the room with Keith Jarret's piano music. To my right, a window—several panes of heavy glass held in place by white steel struts—filled almost the entire wall from the floor to the ceiling. My view outside our Biosphere was always crisscrossed with white steel struts, the bones that supported the glass skin covering most of our world.

From my desk or bed, I could look out the huge window over our half-acre farm below. But it was outdoors with a twist. Beyond lay outside outside. The Arizona desert. The rest of the world. On this evening, a few miles outside my hermetically sealed home, the Catalina Mountains turned a glorious orange under the setting sun.

I heard a soft knock on my door.

Taber MacCallum walked in and plonked himself down on the

sofa. Taber, a veritable bear of a young man only a few months earlier, was now as thin as a rail. Thick brown hair spilled over a prominent brow, under which shone penetrating green eyes. He was attractive, but it was his mind that fascinated me, allured me. The son of an American astrophysicist, he was exceptionally intelligent, thoughtful, and kind.

Taber was my best friend and my lover. I could read him like an open book, and this evening he seemed unusually tense.

"Hi, what's up?" I inquired.

There was a long silence. Finally he broke the quiet. "I'm getting some strange readings in the lab."

"What, the nitrogen generator giving you problems again?" I asked.

"No, it looks like we may be losing oxygen."

Taber was the crewmember in charge of the analytical and other machinery in the lab. We all trusted him to tell us whether our air was safe to breathe, our food was safe to eat, and our water was safe to drink. One of his many jobs was to check daily the level of carbon dioxide in our atmosphere. For this routine chore, Taber made use of the "sniffer system," by which air from throughout the Biosphere was piped into a series of online analyzers in the lab that every fifteen minutes measured the amounts of eight important gases, including carbon dioxide, oxygen, and methane.

We knew from experiments as far back as the days of the Test Module, a sealed chamber one fourhundredth the size of Biosphere 2, that our oxygen and carbon dioxide were locked in a permanent dance. The oxygen level went down in proportion to the amount the carbon dioxide level went up—and vice versa.

Until this day in April, six months into our voluntary enclosure, Taber had tested the level of oxygen only with the sniffer system.

But he had carefully monitored the carbon dioxide level by means of a more precise method called gas chromatography. He knew that there had been a slight increase in carbon dioxide since we sealed the Biosphere, and so it made sense that, according to the sniffer system's analysis, the oxygen level in our closed atmosphere had dropped slightly—to an expected 19.8 percent. This represented a drop of 1.1 percent below the 20.9 percent that is the oxygen level in the terrestrial atmosphere and what we had in the Biosphere in the beginning.

But this time, Taber also ran the oxygen figures from the chromatograph. He was stunned to find data saying the oxygen level in Biosphere 2 had dropped to 17.4 percent.

He did not believe it.

It was impossible that so much oxygen—in all, he calculated, roughly seven tons of it—had simply disappeared. He recalibrated the machine and ran more samples.

Again the computer readout said 17.4 percent.

No way. The oxygen simply could not be that low.

He ran the samples over and over, and the data were unyielding. Only 17.4 percent of our atmosphere was oxygen. This was a significant drop, a horrifying drop, the more so since it was inexplicable.

It meant that our life support system was truly failing. We were about a quarter of the way through our two-year mission and we could have lost about 15 percent of our oxygen. A little more and we'd be in the same situation as mountaineers above fourteen thousand feet—they begin to fall apart mentally and physically if they do not adapt or breathe bottled oxygen.

At the current rate, our atmosphere would be only 10 percent oxygen before the end of the two-year mission: no one could live on so little.

For the better part of two days and nights, Taber sat in the lab amid the cold machinery and blinking lights, his mind churning, desperately trying to understand the enigma.

Something was directly absorbing oxygen—a lot of oxygen—we needed to breathe. Or perhaps it was being used by microbes in respiration to produce carbon dioxide. If that was the case, then something was absorbing carbon dioxide as there was not enough of it in the atmosphere to account for the amount of oxygen missing. In any event, he knew he had to break the news.

After telling me, he went to Abigail Alling, the biospherian who ran the research inside Biosphere 2. Gaie, as we called her, was appalled. She and Taber immediately called outside the Biosphere to the project's director of research and development, John Allen. He refused to believe it.

"Go run it again," he ordered, and Taber obliged, but the three of them finally had to confront the hard fact that a great deal of oxygen had vanished.

Taber spent the entire second night in the lab, trying to come up with an answer. He arrived in the dining room for our morning meeting looking exhausted and pale. The seven of us around the gray granite table all knew by now that there was some issue with the oxygen. We braced ourselves to hear how bad it was.

"Our oxygen is down below seventeen and a half percent," Taber announced in a flat, toneless voice.

The table erupted with questions. What? How can that be? Are you sure? How? Where did it go?

"I have no idea," Taber said. "I didn't trust the instruments at first myself. But I've run so many samples the data has to be right."

The ramifications were immense, and they were not lost on anyone that morning. It wasn't that we risked dying—we could walk out of the airlock at any time if our environment became unlivable. But how could we walk out? At the outset we had declared to the media and the whole world that we would stay inside for two full years.

Leaving Biosphere 2 early was out of the question.

But if we stayed inside, we would likely be forced to pump in oxygen, which would break our promise that no material would go in or out of the hermetically sealed enclosure during our two-year mission.

The media had been hammering us for months, as had the scientific community. Surely, they would both write us off entirely after discovering this latest flaw in the workings of our Biosphere. If the design was flawed, perhaps the whole idea that humans could successfully create an artificial biosphere was also flawed. And if that were true, then Ed Bass, the Texan billionaire who bankrolled the project, would have wasted his $250 million. A quarter of a billion dollars. Perhaps Ed would stop funding us.

And what of the precious "self-organizing" notion we all shared whereby the Biosphere—its overall air, water, life, and chemistry—would seek its own equilibrium and, importantly, an equilibrium that would be habitable by humans. Well, it was self-organizing, all right—organizing us right out of the picture.

We sat stunned into silence. The news was far worse than we had imagined.

I felt a rush as the blood drained from my head down to my feet, and my toes and fingers tingled with adrenaline. My brain was spinning and I heard a voice screaming at me in my head:

"We're screwed! We're screwed!"

Chapter 2

A RARIFIED ATMOSPHERE

S o, how did I, a woman of thirty from a well-to-do family in England, end up sealed inside a giant greenhouse with seven other people, worrying about the complexities of oxygen and carbon dioxide?

It's not as though I grew up dreaming of being one of the first people to live inside an artificial biosphere.

I didn't have a clue what I wanted to be when I finished growing.

In some ways, though, I grew up in another hothouse—the rarified atmosphere of moneyed Britain. The baby girl with two older brothers, I was pampered and protected from the struggle of life. From when I was about six, friends had allowances they managed so, after purchasing a few vitals like books and pencils, they could buy candy and other frivolities.

Not me. My parents didn't even give me pocket money. I simply had to ask for whatever it was I thought I desperately wanted or needed. It wasn't that I always got what I asked for. Far from it.

But I just never had to worry about things like money. I never had to worry about much of anything but sitting up straight and not putting my elbows on the table at dinner.

In the summers my family hung out in the small coastal town of Seaview on the Isle of Wight, off the southern English coast, where it seemed that everyone sailed and knew what everyone else was doing. Occasionally our parents took us on a five-star trip to southern Spain, Italy, or some other exotic place.

My first memory is at two and a half on the beach in Antigua, playing for hours and hours on the beach with "chip-chips," little creatures that looked like pebbles and buried themselves in the sand when disturbed. Returning to Seaview, I spent hours peering under rocks and into rock pools at the crabs and other crustaceans, the snails, fish, and algae. I was fascinated from an early age by the squiggling, crawling, swimming things around me.

When I was eight, we moved into a modern mansion my father had built after knocking down the decrepit Henmead Hall, a Victorian country home that had literally rotted into the damp, green English countryside in Sussex, just over an hour south of London. We watched the brick house grow out of the ground, workmen rubbing gold leaf onto the edges of the cupboard doors in the dressing room. Jim, our gardener, mowed vast green lawns with one of those mini-tractor type lawn mowers. He and his wife Nancy lived in a cute cottage on the grounds, and Nancy helped my mother around the house.

The old woman who lived in their cottage before we took it over was like something out of my fairy-tale books, and the memory of her stuck with me ever after, probably embellished by a child's imagination. She was wizened and bent. I knew absolutely nothing about her, but the day I wandered into her house alone, around ten birds were perched silently along the iron bed head. When I entered they didn't flutter, or squawk, they turned their heads to peer at me. The room was dimly lit and stank from all the bird droppings, but I barely noticed.

The old woman seemed miraculous to me. She had tamed these wild birds, or so I thought—an owl and other various brown feathery things. I saw her as a cross between a witch and St. Francis, living with the animals.

In my house, nature was kept firmly outside the front door, except the odd houseplant and two dogs, a pug and a cocker spaniel. It was spotlessly clean, and everything, everything was always in its place. I had my playroom, and no toys were to be left anywhere other than in the playroom.

But our house was planted near a wood, where I ran free for hours with Dougal, the spaniel, when my brothers were away at boarding school. We romped through the bracken. I climbed trees while he rushed barking after pheasants. I discovered newts in the lake my father had dug in the garden, and one day I found inch-long silvery newt babies huddled in the folds of the lake's plastic liner. I was fascinated, and I looked every day, until they had grown and swum off.

Malcolm, six years my senior, had crammed a small chemistry lab into the attic of an ancient wooden barn that my parents remodeled into a garage. He would drag me up there, and enchant me with chemical gardens. He burned different chemicals to show me the pretty colors the flames made. One afternoon he slunk out of his lab. He had singed the front of his hair, and was missing his eyebrows. My parents closed down his lab, fearing the wooden barn would go up in flames.

In keeping with the social norms of the time, my parents shuffled me off to a single-gender boarding school in the country when I was around nine. But my free spirit got me into trouble. I was all but expelled for having been caught out of bounds (I had wandered off the school property to explore the local countryside) too many times and for having an ink fight with my best friend Clare, during which we flicked ink at each other from our fountain pens. I didn't think it was that heinous, but Clare was told to leave at the end of the term. I was allowed to stay for some reason, probably something to do with my promise not be naughty any more, and specifically not to play with Clare ever again.

When I was twelve, life at home changed radically. We moved out of Henmead and crammed into our holiday home on the Isle of Wight, all five of us and my recently widowed grandmother. I had absolutely no understanding of what had happened, and no one would tell me. They wanted to protect me from the harsh truth that my beloved father, who I hardly saw and barely knew because he was always working, had lost almost everything.

My father, a consummate entrepreneur, was a risk taker. He had made and lost a lot of money more than once. My eldest brother, ten years my senior, had worked with our father. He gave a touching, revealing speech about him at his funeral a couple of years ago. In it he said, "My father was a man of great principle, but if he had one vice, it was his weakness for sports cars. He liked to drive fast, and what a ride we had with him through life."

Somehow my father managed to keep sending me to expensive boarding school, and I graduated to public school, which perversely is the name we English give to private high school. St. Michaels and All Angels School for Girls was a beautiful old stone building, with columns along the front, wisteria blooming profusely purple all over the façade in the spring, and cherry trees lining the long driveway. In the middle of green pastures, rolling wooded hills, a duck pond, it was lovely, really. But it was cold in winter, and we hardy girls scorned those who asked for extra blankets, some as many as seven. I bragged that I only needed the allotted two.

There I met girls from wildly different lives. One girl lived in Kinshasa, Zaire, on the banks of the Congo River, and she told hair-raising stories of gun battles that fired off right in her garden. Another was a close relative of Uganda's Idi Amin. Only a year earlier she had been called into the kitchen at his behest, along with the rest of the royal family. They encircled a table where a dead woman lay. It was her aunt. She told me how the woman's arms were where her legs should be, and her legs had been crudely attached to her shoulders. Amin told the terrified group that this would happen to anyone who crossed him. My friend escaped Uganda with her family a few weeks later.

Then a group of kids came down from some poor neighborhood in London to spend the day at our privileged school. I was flabbergasted when one of the girls turned to us, disgusted, and incredulously pointed at the cows in the nearby pasture, "You mean those dirty animals are where milk comes from?"

I could not reconcile these stories with my life. I began to cringe at what seemed so much pettiness and emptiness around me. By my late teens I felt suffocated by my English heritage, by the senescing biology teacher who would yell at us in her quavering voice to get off the grass, by all those admirals in the yacht club who sounded like they had hot potatoes in their mouths when they talked about nothing much.

To make matters worse, I found life at school entirely unstimulating. Most of the teachers were spinsters with little zest for life. All I had to do on weekends was eat toast, play lacrosse for the school and county, and wander the grounds, contemplating how there had to be more than this. The one teacher who had some spirit left in him was the music teacher, who had written his own opera, in which I played a lead role. I enjoyed the school plays.

So I succumbed to teenage angst.

I was a frustrated, confused, hormone-driven teen. I had inherited my father's proclivity for living life on the edge, and this seemingly idyllic life on the Isle of Wight and the haven of school in Sussex was not it.

But I had no idea what I did want to do, and no one offered me much help. For the national school exams we took at age sixteen I had to write an essay on what I wanted to become. The test paper gave me a choice between an airhostess, a nurse, a teacher, a veterinarian, or married with children.

I was distraught. Why did I have to be any of those things? I had recently met Clare Francis, the Englishwoman who sailed single-handedly across the Atlantic, so I knew that I did not have to take the life I was being offered.

At one of my parents' luncheons I was asked the inevitable by one of their friends, and ostentatiously replied that I wanted to be an astronaut. However, I didn't believe for an instant that it was a real option.

I was starving for life experience, for ideas, for a glimpse of what the rest of the world had to offer. I had had my fill of stultifying schooling, and I refused to go to University. I had to make my escape.

So, with as much emotional energy as the space shuttle has rocket power on liftoff, I launched myself into a life of adventure and discovery as soon as school and the law allowed.

In 1980, at age eighteen, I ran to France on the pretext of wanting to learn to speak French like a Frenchwoman. I accomplished that after six months as a ski bum, making a living working mornings in a grocery. There I met the first real love of my life—a ski instructor who lived in an ancient chalet with no heat in an old Alpine village in the shadow of Le Mont Blanc.

I wandered restless back to England, and I announced to my mother I wanted to go to drama school. "You will not!" came the succinct reply.

Instead, I agreed to go to Pitman's College in London "to have something to fall back on," as my father said. Pitman's was a bilingual secretarial school, where many young women of my station were training to become personal assistants to high-level executives and government employees. The last thing I wanted to be was a personal assistant, but I went anyway, figuring I could temp my way out of England, earning enough money taking temporary assignments to buy a ticket to anywhere.

During lunch breaks I wandered the streets of Bloomsbury in tatty dungarees, not fitting in at all with the other girls in their tweed skirts. On a narrow street just off a small, treelined park behind the college, I noticed a sign hung over a heavy wooden door in a Dickensian London townhouse. The sign pulled me inside for some reason I could never explain.

It read "October Gallery" in simple, red lettering with an abstract black cloud that was vaguely Asian in style, with a red dot in the middle. It looked homemade . . . and exotic.

Outside on a sandwich board was scrawled, "Homemade bread and quiche." This sounded like a good reason to enter, so I opened the door and looked in. The hushed atmosphere completely intimidated me. I saw no one and I backed out hurriedly.

The next day I tried again. I opened the door, took a deep breath and walked into the rest of my life.

A beaming, elegant American woman named Chili, who turned out to be the gallery manager, greeted me. The walls were covered in bold abstract art, as were the walls of the dining room. After a few days of eating lunch without uttering a word to anyone, I finally mustered the courage to strike up a conversation with Chili. She took me around the whole creaking house, and we eventually ended up in the office of the Institute of Ecotechnics in one of the musty rooms above the gallery.

The Institute of Ecotechnics, or IE for short, was affiliated with the gallery, and was involved with projects around the world. Chili showed me fading photographs of intriguing places and took me on an imaginary tour of them.

Aside from October Gallery, there was Synergia Ranch near Santa Fe, New Mexico, where members of IE had built the adobe structures and geodesic domes themselves. They were working on permaculture and irrigation techniques appropriate to the high desert. The Research Vessel *Heraclitus,* a peculiar-looking black boat with big eyes painted on the bow, was sailing up the Amazon River on an ethnobotanical mission.

But Quanbun Downs was what first seized hold of my imagination. I saw pictures of a cattle station in the outback of Australia that was maintaining a working ranch while improving badly degraded savannah grassland. Its sister project, Savannah

Systems, was four hours away along a single-lane highway lined with termite mounds. There in Derby, IE was developing pasture that would be tested at Quanbun.

I saw the beautiful Hotel Vajra in the foothills of the Himalayas in Katmandu. A project in the rainforest of Puerto Rico attempted to grow valuable trees at a profit without clear-cutting the forest. IE had set up a conference center on the French Mediterranean in Aix-en-Provence.

Another project was in the making. The Caravan of Dreams, a performing arts center in downtown Fort Worth, intended to stimulate the social and cultural regrowth of the city center.

Back then, I had no idea what all these projects had in common. The chairman of IE, Mark Nelson, later explained to me the basic tenet of IE. Mark was a thoughtful fellow, who always looked crumpled, even when dressed up in his suit jacket and bolo tie.

Mark said he and several other people, including a man named John Allen, formulated the concept of the Institute of Ecotechnics in the late sixties to reconcile global conflicts between nature and technology—*eco* and *technics*. It was the time of the "back to nature" groundswell in the environmental movement, which rejected the undeniable dependence humans have on high technology.

So the Institute of Ecotechnics, or IE, would take on challenges around the world where technology and ecology could work hand in hand to benefit both humans and the Earth. Cities would be included because, even then, the leaders of IE thought of cities in ecological terms (as people are beginning to do today). IE was founded in 1969, and I was standing in its head office in 1981.

Chili invited me to events. I chatted with artists from around the world. I met Brion Gysin and William Burroughs, and the internationally acclaimed Indian musician, Ravi Shankar—none of whom I had ever heard of before.

She invited me to live there. I immediately moved into one of the small rooms she rented. Along with the room came a whole lifestyle. While I finished up my secretarial course, I learned meditation on Tuesday mornings. I read philosophy on Thursday evenings. At my public school, chapel was mandatory every morning. Now I learned about Zen Buddhism, Confucianism, Sufis, early Christianity. Saturday and Sunday mornings I did theater with Chili and the four or five others who lived at the gallery. Sunday evenings we gave speeches to each other.

It was entirely new, refreshing, challenging, a kaleidoscope of new ideas. Even now, more than twenty years and a lot of disillusionment later, I remember the thrill of seeing new worlds unfold around me. Where others later saw a cult, I found the freedom I had been seeking.

A few months later in August 1982, at the age of twenty, I met John Allen for the first time.

A large, loud American bounded into the October Gallery. Around six feet tall and built like the proverbial brick house, he had a devilish smile, a dimple in his chin, and sparkling, playful eyes. The only thing that gave away his fifty-three years was his graying, thinning hair. He bear-hugged Chili and just about everyone else, including me. Chili had apparently told him about me—the snooty English girl with high spirits, a decent intellect, and an appetite for learning and doing.

John was a graduate in metallurgy from the Colorado School of Mines, with an MBA from Harvard. He was one of those charismatic figures who could be disarmingly charming, or, as I came to learn, cut you down with a glance. Despite a huge presence, he could be as invisible as a pane of glass when he chose to disappear into a crowd. His smooth face could explode with the enthusiasm of a kid at Christmas, or tighten into a sardonic glare that would have made Mephistopheles proud.

A student of history with an acute intellect and phenomenal memory, John thought in historic terms: great battles won and lost, the clashes of civilizations, and the struggle between the metaphysical and the physical realities. I had never met anyone like him.

John had arrived with an entourage of fifteen or so, including Mark Nelson and several others who were to become my friends and colleagues. They were an odd-looking bunch, many sporting berets or fedoras. They had come to the gallery to prepare for the annual tour of their troupe, the Theater of All Possibilities. This year they would be traveling around Europe for a month, before heading over to Canada, the United States, and Mexico.

Somewhere in the middle of it all they were to stop at Les Marronniers, IE's conference center in the south of France, and participate in IE's three-day Galactic Conference, which Mark was going to chair. Apparently, people like Buckminster Fuller were going to be there. John asked me if I wanted to go along. I had finished secretarial school, earned enough money in temporary jobs for the plane ticket. Of course I wanted to go.

My parents, naturally, were incredibly concerned. Not only was I doing exactly what they told me not to do—get involved with drama—but I really couldn't explain what theater had to do with the Institute of Ecotechnics, and what either had to do with meditation, Sunday night speeches, and all the other folderol we got up to. So I stood mute and defiant before them whenever they asked. Slowly, over the next few months, I pieced together the story of IE and the group that would rule the next decade of my life.

To understand what I had unwittingly stepped into, we must take a trip back to Haight Ashbury, San Francisco, in the late 1960s, where a group of white and mostly affluent people was experimenting with counterculture communal living. I wasn't there—I was just starting elementary school in England.

Like many others, they were attempting to find alternatives to the hollow, materialistic values of Western culture. This group included John Allen, Mark Nelson, and several others who would go on to build Biosphere 2. Toward the end of the sixties they wholly rejected the disorganized, drugged-out state of most anarchistic communes, and branched off on their own, eventually settling at Synergia Ranch near Santa Fe, New Mexico.

As a science historian from Columbia University, Rebecca Rieder, said in her thesis on Biosphere 2, "Synergia Ranch became the anti-commune commune."

In Santa Fe the members began to define what an effective group living and working together might look like. They were highly intellectual—many of them with Ivy League backgrounds—and awash in ideas, with Bucky Fuller's Social Synergism a binding concept.

Not only could each individual add to the synergy of the group as a whole, but so did various lines of work, such as theater and business enterprise. Each activity pushed the participant closer to being balanced and reaching his or her full potential than following only a single pursuit.

The wisdom in this approach is echoed in a presentation on rules for leadership, by former U.S. Secretary of State Colin Powell, who writes about the working environment, "Seek people who have some balance in their lives, who are fun to hang out with and who like to laugh (at themselves, too) and who have some non-job priorities which they approach with the same passion they do their work." This outfit was a passionate group of people with several priorities, who generally had a healthy sense of humor.

Life at Synergia Ranch became a creative maelstrom. The group questioned modern life unceasingly, along with humanity's place in it. Out of their trials, errors, research, and effort developed a strict adherence to the simultaneous pursuit of three lines of work: theater to explore inner life, philosophy to search for truth, meaning, and personal development, and business enterprise—initially because they were broke. The Institute of Ecotechnics and its projects eventually became the business enterprise for many in the group.

Meditation, spiritual readings, plays, scientific studies, speech-making all found a slot on the week's crammed schedule and everyone attended each function. The work ethic and discipline were extremely strong, the cornerstone values being "friendship, honor, beauty, and discipline." I attribute some of the efficiency with which Biosphere 2 was built to the high value for discipline and intense focus we all had.

I had already plunged into this life and found it stimulating, exhilarating. But people came and went. Some could not hack the hard work, rigorous discipline, and unrelenting excavation of their inner lives. They left in the middle of the night with a note on the refrigerator. Others completely wigged out. One night, an English painter smeared chocolate cake all over the dining room wall before slinking off.

Where the communes of the 1960s seemed to have little direction and no leadership, this group was extremely can-do and action oriented. They set up a tight structure and leadership that was at times authoritarian. John Allen was the leader almost from the word go.

He would pace before a flip chart on Thursday evenings, drawing expansive diagrams to demonstrate the laws of the enneagram or entropy; lay out the evolution of civilizations; show the geographic, historic, and ideological relationship between the Sufis, Agnostics, and early Christians; or act out how to use "negative" emotion, such as anger or hate, for positive outcomes.

My new life with this crazy, brilliant bunch was like drinking from a geyser. No sooner had we prepared a play about the historic figure Gilgamesh, than we went to Les Marronniers, the conference center. We performed once and then prepared for the IE Galactic Conference.

A dazzling group of presenters graced this gathering, which had no more than a couple of hundred people in the audience. They included Lynn Margulis, coauthor of the Gaia hypothesis, which posits that life changes its environment, making it habitable; Buckminster Fuller, author of *Spaceship Earth,* and perhaps best known as the inventor of the geodesic dome; Albert Hoffman, the Swiss chemist, philosopher, and author who invented LSD; Victor Clube, a top astrophysicist with the Royal Observatory in London; Richard Dawkins, author of the controversial book *The Selfish Gene* and other best sellers; Ornette Coleman, a central figure in what was known as free jazz and recipient of a MacArthur Foundation Genius Award; and Roy Walford, a leader in research on aging and life extension.

It was an eclectic bunch because John, Mark, et al. believed in breaking down the barriers between different disciplines to expand people's worldviews and ways of thinking. Each speaker gave a presentation about some enormous idea, most of which went way over my head. But I did not care. These were the thinkers, the movers and shakers of our culture and science exchanging ideas, and I was hooked. This was the food my soul had been yearning for.

One presentation stood out as both bizarre and significant. Phil Hawes, or TC as he was known, was an architect, a student of Frank Lloyd Wright cum sea captain. He looked the part, with crazy curly locks disciplined only by his captain's hat, and an unabashed grin spread across his swarthy, good-looking features. He was a member of IE. I watched him passionately describe some ludicrous idea about a spacecraft that used adobe or soil as radiation shielding, and life itself as the life support system.

TC had finished describing a pie-in-the-sky idea for a self-sustaining space greenhouse, conceived a year earlier by a small group that included Mark Nelson and John sitting around a coffee table in the outback of Australia, when Bucky Fuller exclaimed, "I didn't think you guys could do it. I didn't think you could design a spacecraft that makes sense, but what you've done here does make sense." He then turned to Mark, John, and the other members of IE who were present at the conference and gave the small band of dreamers the endorsement they needed: "If you guys don't build a biosphere, who will?"

That was that.

A big idea was born before my eyes.

What would become known as Biosphere 2 would be the world's first man-made biosphere, built not in space but on the ground, hermetically sealed off from the rest of the Earth, the first biosphere to be cleaved from the only biosphere in which all life as we know it had existed.

In 1987, Lynn Margulis and Dorion Sagan would write, "There is about biospheres an air of evolutionary inevitability. . . . From an evolutionary perspective what we are witnessing with Biosphere 2 is the budding, the first tentative reproducing of planet Earth as a biological identity. . . . Whereas before we have reproduction of cells and the multicellular collections of cells in the form of individual protoctists, fungi, plants and animals, with the advent of biospheres we now see the first reproduction of ecosystem enclaves as discrete, semi-independent units. This represents something new not only in the limited realm of greenhouses and human technology but also in the larger domain of Earth history."

As the twenty-year-old novice with the group, I had no idea Fuller's comment would determine my life, from marriage to my home, to my eventual career.

Chapter 3

ECOPRENEUR

T he Galactic Conference was over and the members of IE transformed back into actors. We drove off in the theater's bus for the remainder of the tour, sometimes playing to audiences that were smaller than the cast, or using candles to light the stage when the power went off. It was a thoroughly bohemian life—and pretty crappy theater—and I loved it.

The Caravan of Dreams was to open in Fort Worth in eighteen months. This performing arts center would house a jazz and blues nightclub, a theater, and a dance studio. Despite my lack of credentials, John asked me if I wanted to go manage the dance studio. He believed that a good manager could lead any endeavor. Of course I went.

So, after a couple of snow-covered months at Synergia Ranch during the winter of 1982, I flew off to the Caravan of Dreams to run its dance studio under Kathlyn Hoffman, the woman who headed the Theater of All Possibilities, and was the artistic director for the Caravan.

Here I got to know a couple more of the key players in the Biosphere saga, Ed Bass and Margret Augustine.

Ed lived at the Caravan and had been on the theater tour with us. He was an unassuming man, and if you did not know it, you would not have had a clue that he was among the richest individuals on Earth. Ed had a heart of gold and a tremendous sense of duty, and he was practically inseparable from his legal-sized yellow pad covered in Post-It notes. A true Texan, he always wore cowboy boots and blue jeans, perfectly pressed with fold lines right down the center of the pant legs.

Even when I later worked with him drilling boreholes for water and mending fences in Australia and we were both covered in sweat and dust, he somehow looked neat and tidy. He was very easygoing with boyish looks, just one of the lads, and we became close.

Sometime between the theater tour and the Caravan of Dreams he had reached an agreement with IE to fund construction of Biosphere 2 under his organization, Decisions Investment Corporation. John, Margret, and other members of IE would provide the management under an equally decisive organization called Decisions Team. They formed Space Biospheres Venture, quite a mouthful that we all shortened to SBV.

Ed considered himself an "ecopreneur," which was visionary at the time, and his stated rationale for pouring unknown millions of dollars into the Biosphere project was that he would recoup costs from environmental spin-off products. This was a rather charitable notion, as I suspect that he had received hardly a penny from the various venerable IE projects he had already funded, including Quanbun Downs and the Caravan of Dreams.

While at the Caravan, I also got to know Margret Augustine. She would become SBV's chief executive officer, responsible for the management of the Biosphere 2 project through its design, building, and initial operation. She managed the construction of the Caravan, completed at the end of 1983.

Only a couple of blocks from Hell's Half Acre, where cattle drives ended the long trip along the Chisolm Trail (you can imagine the rest), the building maintained its historic façade. But everything behind it had been removed and rebuilt as a state-of-the-art, gorgeous performing arts center, with a striking mural stretching the length of the nightclub depicting the history of jazz and blues.

Margret was one tough Canadian cookie, with a fiery temper, an easy laugh, and a wavy shock of auburn hair that she always pinned back behind one ear. She usually wore large jewelry and pants tucked into a pair of knee-high black leather boots.

She managed with an iron fist. But, one evening after a long week during the Caravan's construction, when we'd had enough intellectual stimulation to last a lifetime, she and I sneaked off together to see the movie *Conan the Barbarian*. She could be very personable, even easy to get along with.

Margret was pregnant with John's child during the last nine months of the construction. Still single, she had a complicated relationship with him.

She was also overseeing the grand opening of the Caravan, an elaborate event that included a homecoming for Ornette Coleman—he and his band played with the Fort Worth Symphony. Less than a week before the event, she had her baby daughter and was right back at it the following day. I marveled at her tenacity.

A geodesic dome housing over three hundred species of cacti and succulents from four deserts around the world stood on the roof of the Caravan—incongruous, but spectacular nonetheless. Tony Burgess, a desert ecologist originally from Fort Worth who would also design the Biosphere 2 Desert, meticulously chose the plants. The display at Fort Worth demonstrated convergent and divergent evolution. There were plants that had evolved independently on different continents that looked almost identical. Then there were closely related species with wildly different appearances. It was an astonishing collection.

I soon realized that the whole theater/dance thing was not what I was cut out to do with my life. When the woman running the cactus dome left, John asked me to take over as its curator, although I had only grown a few vegetables. I read books, consulted Tony, and the plants survived.

When the woman running the public-relations department also quit, John asked me to take her job, too. I found myself driving blues singer Koko Taylor in her beat-up van to radio interviews, and hanging out with Ornette Coleman, Carmen McCray, T-Bone Burnett, and so many other great artists. It was exhilarating.

I was a quick study, could manage, and had a penchant for diving up to my neck into things I knew little about and swimming anyway. However, I will never know for certain what qualities I had that made John sure I could do the things he asked me to do. He belonged to the school of negative feedback. He rarely gave positive strokes. Once I asked him, "You never tell me what I do right. How will I know I'm doing a good job?"

"Don't worry, kiddo," he responded, "You'll know because you won't get any feedback."

News of Biosphere 2 wafted over to the Caravan from time to time. When the members of IE got together and decided to do something, they didn't sit on their rear ends and think about it interminably, they got on and did it. By December 1984, two years after the Galactic Conference, Mark, John, TC, and an engineer member of IE named William Dempster had drawn up initial concepts. They had also bought land twenty miles north of Tucson, Arizona, on which sat a conference center that was previously owned by Motorola. That winter IE held its first conference on the future site of Biosphere 2.

Margret, John, and Ed had decided that southern Arizona was one of the best places for the project because it boasts an annual average of well over three hundred days of bright sunshine, important for a biosphere that would rely on natural sunlight for plant growth. The structure would not have to withstand earthquakes, which rarely occur there. The University of Arizona, close by in Tucson, would serve as a key resource in the design and building of the Biosphere, particularly its Environmental Research Laboratory and Arid Lands Institute. Other experts also lived nearby.

I had left Cowtown and life in a performing arts center that ill-suited me, and was heading to Quanbun Downs and then the R.V. *Heraclitus* after a month's theater tour in Nigeria. On my way, I joined the hundred or so attendees at the SunSpace Ranch Conference Center in Oracle, Arizona for a three-day conference devoted to the Biosphere.

The conference center sat on a ridge, overlooking a deep canyon leading all the way to Tucson. Buffalo Bill was rumored to have lived in a cabin somewhere along the canyon. Coyotes and wild pig-like critters called *javelina* were common, and the occasional gila monster and rattlesnake sunned themselves on the black tarmac road. It was a glorious setting, and the night sky burst with stars.

The conference opened at SunSpace Ranch on December 7, 1984, with John delivering a heady speech about how much Biosphere 2 would contribute to the future of mankind. Mark Nelson provided an equally historic perspective, saying, "Space Exploration has given us access to the physical world outside the Earth. To create Biosphere 2 will give man his first opportunity to step into a new living world."

Such was the excitement and reverence with which those involved approached the project. Arrogant yes, but the world's media were not far behind in chiming in with similar accolades.

It was a remarkable three days of early concept reviews and other topics ranging from the role of microbes in biosphere equilibrium, to places to go in the universe once we had a biosphere to travel with. The culmination was a talk by Rusty Schweickart, an astronaut on *Apollo 9,* about the realities of life in space. Through it all, we never suffered a shred of doubt that Biosphere 2 would be built.

Chapter 4

SEA CHANGE

T he bow gently rocked up and down, lazy, hypnotic. The cobalt-blue Indian Ocean rolled away to the horizon in every direction. A red-footed booby stood on the edge of the crow's nest atop the eighty-foot-tall main mast. The bird must have flown a long way, as we had already been at sea for more than two weeks. The wind had died shortly after we left Sri Lanka, where, in October 1985, I had first stepped aboard the Research Vessel *Heraclitus*.

We were not traveling fast, five to six knots at most, maximum speed under motor. We would reach Port Said in Egypt in about six months. I would disembark there, while the ship continued on to Puerto Rico, and ultimately an expedition to the Antarctic.

Now we were headed southwest past the Maldives, an island nation that will probably disappear under the ocean when global warming raises sea levels during this century.

The R. V. *Heraclitus* was a ferro-cement (concrete strengthened with iron re-bar) research vessel that IE staffers had built themselves, re-bar by re-bar, in the early 1970s. Several developing nations were then experimenting with seagoing fishing boats of ferro-cement, a material far cheaper and maintainable, they hoped, than metal or wood. For safety's sake, she was built like a Chinese junk, with a wide beam, three masts, and junk-style battened rigging. Painted black and red, she had two huge eyes painted on the bow, a tradition common to Chinese, Egyptian, and Greek seafarers. The eyes helped her find her way safely across the seas. At eighty-four feet long, she officially qualified as a ship, not a mere yacht.

I sat on the bowsprit, looking down at the reflection of the black hull in the water. My naked legs dangled loosely on either side as I watched the bow slide down into the water and slowly rise out again in endless slow motion. The white bow wave caressed my mind, washing away extraneous thoughts.

The sea had been so calm that we turned the engine off for half an hour each afternoon to swim in the cool water. I heard the engine die, and the deckhand called everyone to leap overboard, save the two on watch.

The water was thousands of feet deep, but I was not afraid that some monster would come from the blue depths. We had seen no marine animals except a few hapless flying fish since leaving Sri Lanka. But today, as soon as we were in the water, a huge shadow appeared below us.

My heart leapt into my mouth.

The amorphous shadow took shape as an enormous, regal manta ray. It swam slowly and elegantly toward us, then swam with us. Its black-and-white wingspan must have been fifteen feet. Gently it swept its wings up and down, gliding round and round, under the ship and through our group as we hung, suspended in the water. The huge animal grew ever closer until finally a couple of us gently held on as it pulled us around and around.

We were surely breaking some rule of marine-animal encounters that none of us thought about at the time. Finally, it was time for us to climb aboard the ship and move on. As soon as the last person left the water, the ray vanished.

Sometimes we need extraordinary experiences to shock our senses alive, to rip a hole in our well-protected worldview. Swimming with the manta ray was one of those experiences. For me, this was not simply an extraordinary encounter with a curious creature that emerged from the deep. The animal had reached into my mind and said, "Hey, human, there is a whole wild world down here, with majesty, intelligence, and beauty, a place where you guys are not at the epicenter. You are merely visitors here."

After growing up coddled in pastoral England, where every square inch has been shaped by *Homo sapiens* for our benefit, I for the first time awoke to the intertwined biosphere in which we live, and to the realization that we are but a cog in the great wheel of life here on Spaceship Earth.

Although I did not yet know I wanted to go inside Biosphere 2, this was the biospherian training program. The R.V. *Heraclitus* and Quanbun Downs, the cattle station in the Australian outback, were mandatory stops along the way for each candidate to become a biospherian. Isolated, with small teams relying on each other for prolonged periods, John and Margret considered the ship and the station close analogues to being enclosed in Biosphere 2. They were biospherian boot camp, sometimes extremely harsh, sometimes magical, intended to push us to or beyond the limits of our ingenuity, alacrity, and imagination.

Our training program could hardly have been farther from NASA; highly structured onslaught of centrifuge rides, mission simulations, medical tests—all intended to prepare the astronaut for every rattling, g-pulling, or floating sensation, every possible problem or crisis that could occur. Our training in the outback and on the *Heraclitus* showed none of the outward signs of order and structure that mark the astronaut program, but it attempted to prepare us for the one aspect of the whole project that we could not learn at the project site—working in small groups in isolation.

Now that astronauts have lived onboard the International and Russian Space Stations for months at a time, psychologists believe that the core of "the right stuff"—the impenetrability of the astronaut's psyche—does not help one spend a long time in the isolation of space. The bone-hard astronaut exterior, the "nothing's going to faze me" attitude, so extolled during the *Apollo* and early days of space shuttle flight, does not bend. When pushed too far, it breaks.

Those who have done the best under the endless deluge of stressors are those who can bend with, and rebound from, lapses into despair and anger, month after month. Instead of resistant, the new astronaut needed to be resilient. As unusual as our training was, it selected for resilience.

As it turned out, every one of us on the *Heraclitus* was a potential candidate to live in the new world. John and Margret observed us from afar every step of the way.

The evening after the manta encounter, I was back at my favorite spot on the bowsprit, gazing at the setting tropical sun as it transformed the ocean into molten vermilion and sank rapidly below the horizon. I sought but did not see the elusive green flash—a rare phenomenon caused by the earth's atmosphere refracting the sun's rays, painting the sky emerald for a moment just as the sun's disc disappears from view. The first stars were barely visible in the fading light. I had never been anywhere so tranquil, so peaceful. Is this what it would be like, I wondered, living in isolation for two years?

Eight rings of the ship's bell broke my reverie.

Taber MacCallum, the second mate, yelled from the main deck: "Hey, HQ! It's time for our deck watch." Most of the crew and members of IE had taken nicknames and mine was Harlequin, or HQ for short, I suppose because I was something of a ham, and could cut a pretty good rug when called upon to do so. I welcomed the name change because "Plain Jane with no nonsense," as my mother would joke, was not exactly my idea of an identity. HQ stuck with me even while I was in the Biosphere.

Unbeknownst to me, the winter-over crews at the South Pole also take nicknames. This, it seems, is an acknowledgement that the experience has transformed them. It's made them different from all those who have never experienced the particular solitude that comes with extended isolation with a small group, and the personal release that follows when the insanity of our impression-packed modern life is stripped away. One is left staring into one's own unfamiliar face, the eyes speaking of things one had never known lay behind them. On the ocean it is called the sea change, a term first used by Ariel in Shakespeare's *The Tempest.*

Like a cake cannot be un-baked, a person cannot undo the sea change, and the watery world never quite looks the same again. The solitude and the ocean chip away at the landlubber's view of planet Earth as continents, little chunks of land separated by barriers of water and national pride. Instead, the sea becomes a welcome constant, encircling planet Ocean uninterrupted.

Taber had no need of a nickname—Taber, his name since birth, was already uncommon enough. "On my way," I called, turning to look back at the impossibly enthusiastic expression on his youthful face—the all-American boy, I was sure, with his bronzed torso and baggy shorts. There was an innocence in his earnestness. His eyes displayed a calmness, a depth that suggested a lifetime of experience. He was compact, made like a football player, but it had been soccer that took out his knee.

"I'll take the helm, if you like." I offered. Taber agreed, and went below to finish overhauling the diving gear for our arrival at the Seychelles. He was the dive master on board, being a diving instructor and licensed commercial diver. He left Ibis on deck watch; the lanky, soft-spoken German was the third member of our dog watch. We were on from four to eight, morning and evening.

Because the ship was always moving, in need of attention twenty-four hours a day, we held watches like all ships at sea, four hours on watch, eight hours off. We rotated watches every couple of weeks. I loved the gentleman's watch, eight to twelve.

The graveyard was the hardest, twelve to four. After being on the helm for an hour at three in the morning, concentrating on the little line showing our compass bearing, wanting nothing more than sleep, I'd keep myself awake by focusing on that line until it was seared into my retina. When I'd look up and out at the black night, I would still see the line swinging before my eyes.

How I hated the graveyard watch. But I needed the prissiness of my five-star hotel upbringing squeezed out of me. Graveyard taught me to keep alert and working even when dog tired. We'd do a lot of that in the Biosphere.

During night watch, when we could not see enough to work on ship maintenance, we'd talk about all kinds of things to pass the time and keep ourselves awake. This evening, when Taber was done with the diving gear, we chatted about his work with Dr. Clair Folsome of the University of Hawaii, a historic figure in the Biosphere lineage.

Folsome was the first scientist to seal life in a jar for prolonged periods. In 1968, he started with material close at hand—Hawaiian beach sand, with the bacteria and algae that inhabited it. He poured the sand into flasks, sealed them, and proceeded to try to kill the life inside. He shook them, froze them, kept them in the dark.

But, no matter what he did to the flasks, unless he heated them to the point that the proteins broke down irreparably, or deprived the systems of light for a long time, life always persisted inside, making pink, orange, and green striations through the sand.

He concluded that the fluctuation of the bacteria and algae species within the ecosystem kept the environment fit for habitation. In short, life alters its environment for its own benefit.

This was a microcosmic embodiment of the Gaia hypothesis, which postulates that, "the physical and chemical condition of the surface of the Earth, of the atmosphere, and of the oceans has been and is actively made fit and comfortable by the presence of life itself." The Gaia hypothesis and Folsome's flasks ran counter to the conventional wisdom that life adapted to the planetary conditions it found. Some of the Folsome systems continue thriving after thirty years spent hermetically sealed.

According to Taber, at the time of our voyage, Clair was attempting to answer a vexing question for Space Biospheres Venture, the Biosphere 2 project. In order to have a fully functioning mini-rainforest, or a mini-savannah, did the soil and microbes have to come from a natural savannah, which would mean going as far afield as Africa or Australia, or would local soils with appropriate physical properties and local microbes do the trick? It seems so obvious to us today, but twenty years ago the ecology of microbes was still a fledgling science.

To aid Clair's quest, Taber was collecting microbial mats and soups from different seas, depths, and ecosystems around the world in little white vials, and sending them off to be analyzed. Collection entailed scraping little pieces of coral reef gunge, hull goo, and other tiny bits of sea stuff off their various substrates, and into the small opening of the vial—a tough task with even the slightest water movement.

Using these samples, Clair discovered that no matter what location in the ocean, whether near the surface or sixty feet down, on a coral reef or in the open ocean, if there was nitrite, then there would be one or more species of bacteria turning it into nitrate that algae could use.

From a biosphere designer's point of view, it did not matter what particular species did this important job. With Taber's help, Clair saved SBV a great deal of money. Assuming that the same would hold true in soil, they could use local soils with local bugs.

Taber, two years my junior, was ahead of me in the race to be one of the first eight people to be sealed inside Biosphere 2. In 1984 he had been initiated into the training in a rite of passage of sorts, one that he and his fellow *Heraclitus* crewmembers barely survived.

Taber grew up in Albuquerque, New Mexico, an environment very different from my upbringing. He had three grown half-brothers from his father's first marriage. He also had a younger full-brother, Ari. Like his older brothers, he called both of his parents by their first names, Anthea and Crawford, which I always found odd. Anthea was an artist and therapist. Crawford was an astrophysicist. Crawford started taking Taber to professional scientific conferences when he was only ten years old. Even then, Taber felt at ease talking with astrophysicists and astronomers about the wonders of the universe.

But when he was twelve, his parents went through an ugly divorce, and Taber was often left alone to fend for himself and Ari.

He found little solace at his urban high school. It was windowless, with razor wire atop an eight-foot fence. Police patrolled the grounds and hallways. A student in Taber's class once threw a desk across the room because he didn't like a grade the teacher had given him, hitting and almost killing the teacher. This was not a particularly unusual event. Taber languished.

When he was fourteen Taber fled to Austin, Texas, against his mother's wishes. A linguist and concert pianist, who was teaching summer school at the University of New Mexico, had rented a room in Crawford's house for a few months earlier in the year. After seeing Taber's depressing situation, he invited Taber to go home with him to complete school at Austin High, which had an excellent reputation. Taber gladly agreed. The man and his wife became his legal guardians, and welcomed him into their home. Taber thrived, graduating with good grades and thoughts of going to Harvard.

But he wanted to see the world before entering college, so he gave himself three to four years to travel. Where I had had no plan upon leaving England, Taber had big plans. Having been raised during the Cold War, he wanted to visit what he considered the other world powers—the U.S.S.R., China, Europe, and Japan.

In 1982, after saving enough money working as a car mechanic, the eighteen-year-old flew to Europe. During his nine months there, he visited the October Gallery, which he had heard about from one of his older brothers, who then lived at Synergia Ranch.

Chili greeted Taber, as she had me, and graciously showed him around the creaking building. He saw the same fading photographs of all the IE projects that had grabbed my attention. The photo of the strange black ship with the three masts transfixed him. Having lived in the desert all his short life, he was dying to go to sea. Chili told him that the R.V. *Heraclitus* was sailing around the world. She would dock in Singapore about the time he was planning to be in China—a mere hop, skip, and a jump away. Furthermore, IE ran an onboard program whereby any able-bodied person could apprentice as a seaman. Taber signed up for a six-month program he would later join, and headed off to explore more of Europe.

Three months later, he took the Trans-Siberian railroad from Moscow to Vladovostok. Contrary to what he heard from American propaganda, the Soviets he met felt secure. They always had a job, always had health care. But the price was personal freedom.

They were convinced that in the U.S., people died from no health care. They imagined thousands of people killing each other because of extreme racism and unemployment. To them, America was a terrifying place.

During his nine months in Japan, he never understood Japanese thinking. How was it that on a certain day of the year, every office worker began wearing short-sleeve shirts instead of long sleeves? To Taber, watching Japanese culture unfold was like gazing at a school of fish that seem to magically move together as one.

In China he witnessed oceans of Mao-blue suits and black bicycles that seemed to go on forever, and what appeared to him to be a kangaroo court. Two men were standing trial for he knew not what. Uniformed men plastered cartoons of the crime scene on the walls. As far as Taber could tell, the men standing trial received no defense. They were executed.

This trip transformed his previously jaded view of America. He appreciated his right to choose, his right to free speech, and his right to a fair trial. He understood that America was far from perfect, but after what he'd seen, he felt downright patriotic. After *Heraclitus,* he would be ready to go home, to go to Harvard.

For reasons that the ship's captain had not made clear to Taber, the *Heraclitus* had not made it as far as Singapore. Instead, he flew into American Samoa in the South Pacific, where he was to take a short hop to a neighboring group of islands and embark on the ship. While waiting for his plane he chatted with a couple of locals in the bar. Upon hearing that he was on his way to join up with a ship in Western Samoa, one local turned to the other and said, "Isn't that the boat they're going to dynamite off the reef next week?"

This was Taber's first inkling of bad news. John Allen, who had come to inspect the ship's damage, soon confirmed that the *Heraclitus* had indeed been reefed at the entrance to the port of Asau. The crewmember piloting the boat through the channel into the harbor had wandered off course, grounding the ship on a reef on a falling tide. The razor-sharp coral smashed so many holes in her ferro-cement hull that she sank onto the reef, filling with water and emptying with each subsequent tide.

John said he would buy Taber a ticket to anywhere in the world, but Taber had not flown so far only to turn back. He decided to stay and help rebuild the ship and sail on when she continued her voyage around the world. John also told Taber about Biosphere 2. Taber was intrigued. But he didn't give up on his goal of attending Harvard.

When Taber reached the *Heraclitus,* the crew had epoxied temporary steel plate patches on the hull. The crew finally floated the ship off the reef and into the port of Asau. The seven-person crew, including Taber, then spent grueling months welding re-bar, mixing cement, patching more holes, and repairing water damage throughout the ship.

Spending American dollars, they were most welcome on the small island of Savai'i. It was far more money than the islanders were used to having infused into their economy.

Unfortunately, the island's officials were not about to let their cash cow simply sail away. After the crew refused to marry into the local community, the ship's captain got wind that the harbormaster was coming from an adjoining island to impound the ship on a trumped-up charge of bad debt. It was Good Friday and he would not be arriving at the ship until Monday. The holiday gave the crew three days to pack up and ship out, but the ship was entirely empty save the engine and masts.

They hired every able-bodied person trusted by a friendly local chief. They loaded supplies and food, fitted floors and bunks, rerigged the masts, installed the galley, and finished whatever other last-minute refurbishments they could before they set sail due west as the sun rose on Monday morning, a full six weeks earlier than scheduled.

Shortly after they left the harbor, the metal plates patching the hull started flexing and the ship began to leak. During the ensuing thirty-six-day voyage, she took on fifteen tons of water a day. When the generator broke down, the crew pumped the bilges continually by hand. She lost her engine and steering. Diesel contaminated the supply of fresh water, so the crew collected rainwater on the sails. They relied on celestial navigation, but the sky was cloudy for days at a time.

Finally, they lost all radio communication and those of us in the U.S. feared the ship might be lost at sea along with her crew. Someone at IE sent an airplane to search, but finding a small boat in the midst of the Pacific was like looking for a particularly small needle in a vast haystack. I was at the Caravan of Dreams in Fort Worth, tending my cacti, when the news came that she could not be found.

Through luck, ingenuity, and superhuman effort, they made landfall in the Island of Vanuatu. The local men, dressed in long sarongs they call *lava lavas,* greeted them with baguettes and fresh butter. The island had been a French colony and baguettes were flown in fresh daily.

Taber had certainly earned a position on the biospherian candidate team. He decided that an opportunity like the Biosphere 2 project comes around once in a lifetime, at best. Harvard would always be there if he wanted to go later. He applied to be a biospherian candidate, and John accepted him.

Mark "Laser" Van Thillo had been through the same grueling voyage. He, too, was now a candidate. Laser, the ship's Belgian chief engineer, was so tall and thin he could have slipped through a drinking straw. Taber and Laser developed a strong bond during this trial, and subsequent *Heraclitus* adventures.

I, on the other hand, was not yet a candidate, but had stumbled into the training program on my tour around the IE projects. I was twenty-two, and did not understand the significance of Biosphere 2, but it sounded intriguing. I became caught up in Taber and Laser's excitement about it.

Later, people would ask me why I wanted to give up two years of my life to go inside Biosphere 2. I could never understand this question. I did not view it as giving up two years, but gaining them. I wanted to be part of something bigger than me.

It was historic. It would make my career. It would be the closest thing to living on Mars. I would find out for myself whether man-made biospheres work. I would experience being enclosed for two years, isolated from the world, so I could impart my knowledge and experience to those who followed, hopefully on their way to Mars.

It would be transforming. None of us who went into the Biosphere would be the same people when we walked out again—a hero's journey, in the Joseph Campbell sense.

But back then on the *Heraclitus,* I still had to earn my place on the team. I had to prove my character, my resourcefulness, and whatever else it took. I had no idea what qualifications were required, but I would apply once I arrived in Egypt. Until then, we had the Indian Ocean to cross and the Pirate's Red Sea to sail.

After two weeks of motoring toward the Seychelles, the sound of the engine was becoming tiresome, and we had all but emptied one of the ship's two tanks of diesel. When Laser went to turn the engine over to the second tank, the valve malfunctioned, and we could not access the fuel. If we ran out, we would be becalmed in the middle of the Indian Ocean. Fortunately, we were close to a grouping of islands in British territorial waters called the Chagos Archipelago. One of them, Diego Garcia, hosts a U.S. naval base, and was smack dab in the middle of our nearly two-thousand-mile journey to the Seychelles.

We radioed the base for help. The officer replied that if we declared an emergency, then they would have to allow us entry. So we declared the emergency, and our black and red Chinese junk docked alongside gray American ships of war. The base crew lavished hospitality on us. The Navy had ice cream that had expired, so we helped them dispose of it. We ate what we could not fit in the freezer, and for a day we rolled around on deck holding our stomachs from sugar poisoning.

After repairing the damaged fuel tank, we thanked them heartily, then spent a couple of blissful days at an uninhabited island, anchored in a classic bay with white sand and palm trees. Sea turtles swam in the bay and starfish were everywhere, preparing for the annual mass mating when they would collectively release billions of eggs and sperm into the bay's calm waters. We caught jack fish for dinner. Enormous land crabs on shore, bigger than a coconut, stared out from hollows of tree roots. It was like living in one of those Jacques Cousteau movies I loved so much as a teenager. Here, in this paradise, I learned to dive.

Diving is a skill that I and several other biospherians would use extensively during the building and the two-year closure of Biosphere 2. Aside from enabling work underwater during trips to collect organisms to stock the Biosphere 2 Ocean and in the mini-ocean itself, diving taught us vivid lessons about atmospheric control, and to think about the medical ramifications of blood chemistry. You can kill yourself quickly if you do not handle your equipment safely, fill the bottles correctly, have the right mix of gases, and allow the nitrogen to diffuse out of your blood stream slowly as you return to the surface from a deep dive.

We dived at every opportunity. In the Seychelles, I saw my first octopus, a rare red species that danced across the sand in the dark water of night, changing colors as it moved. And now we were headed to the Red Sea, one of the world's greatest diving centers. The wind began to pick up and soon we were sailing, the noise of the engine gone, replaced by the wind in the rigging, and the ship crashing over waves.

We took turns cooking the day's meals on a wood-burning stove, particularly tricky in a big sea. One evening Taber was deep-frying chicken. I walked behind him to get a drink of water just as the ship hit a big wave and the pan flew off, pouring boiling oil down the back of my legs. Taber was devastated, though it certainly was not his fault.

I learned to trust my fellow crewmembers then. We had no doctor on board, though Captain Rio, a.k.a. Robert Hahn, had training in first aid. We would not reach port for two weeks, plenty of time for me to incubate a dangerous infection. Twice a day, Rio changed my dressing and spread antibiotic lotion on huge blisters and bloody flesh that covered both calves. As the wounds began to heal, he spread vitamin E oil over them, saying it would help reduce scarring. By the time we got to Yemen, the burns had almost healed and would leave no scar.

Meanwhile, I continued my responsibilities on watch. Taber and Ibis rigged a rope so I could suspend one leg at a time in the air to reduce the pain while on the helm. This experience gave me confidence that the members of small groups surviving in challenging conditions keep each other safe from serious harm. This would be put to the test when I cut off part of a finger in Biosphere 2.

Finally, more than two months after leaving Sri Lanka, we

arrived in Aden, South Yemen. Many aspects of life at sea were analogous to life inside Biosphere 2: a small crew together in a cramped, isolated environment, learning to depend on each other, to work together as a team, to build and tend to relationships even when your crewmates got on your nerves, to live with fewer sensory inputs, to perform hard manual labor, to live in nature and not separate from it, to learn flexibility and ingenuity.

But going into port, the analogy broke down. As any ocean traveler knows, port is a place to blow off steam. We would not have that luxury in Biosphere 2.

There would be no outlet to reduce the pressure that builds up among eight people breathing down each other's necks while suffering sensory deprivation. Isolation of more than four to six months causes social and psychological phenomena that one simply could not prepare for on thirty-day voyages. We of course did not know this. We expected to be totally ready for life on the inside.

Control freaks do not last in sealed biospheres, or at sea. The ocean teaches you to relinquish control to forces greater than you. On calm days in the Red Sea, dolphins often swam at our bow, leaving trails of green phosphorescence at night. Then a storm would blow in, transforming the tranquility to angry waves flooding the decks.

During a particularly bad storm, we were blown backwards. After battling on for hours, sure we could somehow make headway, even risking damaging the rigging, we reluctantly backtracked and anchored in the lee of a natural harbor, a horseshoe atoll. As we entered the bay, dolphins leaped at the bow, leading the way to safety.

Taber, Laser, and I became close friends and diving buddies. I grew particularly fond of Laser on this voyage. I would see him disappear into the cramped engine room to nurture the belching diesel engine. When he emerged hours later, his lanky body was smeared with black grease. When he cleaned up, he donned a black long-sleeve shirt over black jeans. Both were faded, but they made him look somehow casually elegant. He swept back his short, straight, dark brown hair to reveal a long face often attacked by outbreaks of acne. But he carried himself like a stallion, his head high and his chest ever so slightly puffed out. Full of self-assurance, he was headstrong and emotional. He flew into a rage on a hair trigger, but not at anything petty. Then the storm would pass, and he'd laugh his chesty laugh.

The three of us dived to experience a coral-reef ecosystem like that in the mini-ocean planned for Biosphere 2. We dived off a lighthouse far from land that sat atop a coral atoll. Its walls dropped straight into the depths, a coral cliff face with an astounding diversity of reef life. Huge moray eels poked their heads out from their hiding places amidst clouds of delicate butterfly fish and silvery surgeonfish. Here we learned the astonishing beauty and complexity of the natural world. Here, we not only bonded with each other, but in some sense with the creatures we met, with the wilderness as yet untamed or damaged by humans.

We dived to fill white vials. During one trip, Taber was nearly eaten by a killer Napoleon fish, a six-foot, emerald green fellow with a gnarly lobe hanging down its face. The fish swallowed Taber's arm up to his elbow, slowly crunching down with razor sharp teeth that began to pierce his wet suit. Finally the fish let go of his arm but kept the white vial Taber held in his hand. Eventually it spat out the vial but continued to pursue him. Horrified, Taber and his diving partner Laser pulled themselves out of the water, and shakily recounted the event to local bystanders—who collapsed laughing. Apparently the fish was so large because it had been fed hardboiled eggs by the locals. It must have thought the vial was a particularly tough egg.

We dived to visit history. Sha'ab Rumi atoll was the site of one of Jacques Cousteau's daring experiments in underwater living. In 1963 Cousteau and his "oceanauts" lived in the marine village called Precontinent II for six weeks. When we explored it, only fish cages remained, and a submarine garage that looked like a toad-stool. We swam inside and found air still trapped at the top of the garage, just enough for us to take off our masks and have our first underwater conversation. I imagined underwater biospheres, cities where people lived not for weeks, but for their entire lives.

In keeping with the norm on all IE projects, the entire crew, save those on watch, performed forty-five minutes of meditation twice a week to learn to focus our minds. John also considered theater a great way to explore one's emotional and subconscious life, vital for anyone about to be locked up for two years. Because of my previous theater experience, John had designated me the onboard director. The fourteen-member crew wrote skits and most of the songs of a play that we performed to a packed house in Djibouti in an outdoor amphitheater.

In the middle of our rendition of Rimbaud's "Le Bateau Ivre," stones began showering down onto the stage. For a terrible moment we thought the audience despised us so much they were stoning us. On the contrary, a group outside had not been allowed in and were so upset that they were chucking stones over the wall in protest. None of us broke character and, following my lead, we went on as if nothing had happened.

On the Red Sea I wrote my first poem. I slept on the deck at night and inhaled the starscape. We dived on the most staggeringly rich coral reefs in the world. A Montenegrin prince was filming on board the ship and was as expressive as the Englishmen I had known were undemonstrative. He opened my eyes to the beauty of different lighting. In his thick, Yugoslav accent dripping with passion, with his shoulder-length black hair blowing, he would demand, "look at dis light, just look at dis light! It so beeootiful! It's de bess!" And he was right, the warmth and depth of the evening sun was and still is the best.

This was learning to live. This was learning to breathe deeply the heady scent of the roses that grow along our path, to be in the present moment, appreciating what is around me instead of racing blindly through life. What I did not know was that this simple love of life in the slow lane, something traditional people have mastered and that is anathema to our modern life, was exactly what we needed to survive being voluntarily sealed away from the world for two years. This was learning to be a biospherian.

We finally arrived in Port Said. I exchanged telexes with John, who accepted my application to be a biospherian candidate. After performing a cabaret we had written for the chief of police, I said *au revoir* to Laser and Taber, who sailed to Puerto Rico. I fervently hoped to see them in Arizona the following year. I left the sea behind me and flew to Australia for my second stint at Quanbun Downs in the outback, one of the least populated areas in the world.

Chapter 5

EXTREME TRAINING

I first saw Quanbun Downs in 1985, just before the wet season. The ground was so parched that nothing but a few trees grew on the baked clay flats. A gray and desiccated plain stretched for miles across the Kimberley, an area of nearly two hundred thousand square miles that even today has fewer people per square mile than almost anywhere else in the world. In the areas of red sand, patches of inedible olive-green, prickly spinnifex grew around sculpted termite mounds.

I was astonished that cattle lived there at all. In the hot dry, it is 120 degrees in the shade. In the wet, it floods. The outback of northwestern Australia truly is Alfred Lord Tennyson's "Nature, red in tooth and claw." It's all about survival.

John and Margret considered Quanbun an excellent analog to life on the inside. I would work there for twelve months between 1985 and 1987. Extremely isolated, the station (as cattle ranches are called in Australia) was small by Australian standards, a mere three hundred thousand acres, tended most of the year by a tiny crew of between three and five people. The ground had been horribly desertified by sheep grazing, though some of the pastures were now much improved by the heroic efforts of the station manager and the people at Savannah Systems, our sister project that propagated grass seed.

The nearest settlement, Fitzroy Crossing, was a one-and-a-half-hour drive down rutted, sand-trapped, potholed dirt tracks that became flooded and impassable during the wet season. Fitzroy at that time had a pub, a small general store that held the post office, a tiny clinic, and an Aboriginal settlement.

That was it.

For anything more, we drove another four hours to Derby on the coast. Nowhere was a Home Depot to be found. If something broke, we had to fix it ourselves with what we had on hand. It was basic training.

As Silke "Safari" Schneider, another biospherian candidate, later said, "A lot of people have never experienced any type of remoteness. There were a lot of people who found that very challenging. Some could not stand it and left." Between life on the heaving ocean and in the baked savannah, all biospherians underwent training in extremes. These were our rites of passage. I was at Quanbun longer than most, but everyone got a taste of what I experienced.

The outback seemed caught in a time warp of one hundred years ago. Quanbun was sandwiched in between a station owned by "the Texan Boys," three brothers nicknamed after their home state in the U.S. who looked and sounded like they had stepped out of an old cowboy movie, and a station owned by "Uncle Bob," the rustler. The two stations were locked in a long-running feud. The Texan Boys claimed Uncle Bob was stealing their cattle.

Sometimes when we were sitting having dinner at the homestead, we could hear a helicopter flying over Quanbun down by the river a few miles away. Diana Matthewson, the station manager, would say quite matter of factly, "There goes Uncle Bob, running our cattle across the fence onto his land, again." The Texan Boys were probably right.

I never met Uncle Bob, but I heard that shortly after I left, the Kimberley law got onto him, and he fled to Queensland to race greyhounds.

I had a love-hate relationship with Quanbun. The people were colorful, life's agenda simple, and the facilities basic. Most of the buildings sat on stilts, with no solid windows, just slats to allow whatever breeze wafted through to help relieve the stifling heat.

I lived in a small room with a tin roof that was an oven during the day. The bed legs rested in coffee cans filled with water so ants would not crawl up and under the mosquito netting that hung around the bed. A blue-tongued skink lived under my doormat—a foot-long lump of lizard that I had to take care not to step on.

In the women's bathroom, I became strangely fond of the huge green frogs that sat and stared at me from inside the basin or bathtub. The boundary between nature and our human habitat was blurred. Occasionally, if one did not take a good look in the toilet before taking the throne, a surprised frog would jump up out of the water and grab the shrieking person's backside.

When I arrived in late March, all the seasonal ponds called *bill-abongs* were dry. Our crew of five struggled daily to help cattle that were scattered across hundreds of thousands of acres survive the dry heat and the lack of water until the wet season arrived a few weeks later.

Ed had come to the station from Fort Worth for the roundup season, which was starting soon. I would drive out with him and Diana in aging Land Rovers with failing brakes or loose steering. We bumped along dirt tracks to check water troughs, fix well pumps, and throw hay to skinny cattle that lay around the watering holes. I learned to dig out of sand when the vehicle got stuck up to the axle, and fix holes in fences that we came across.

Along the way we passed skeletons of cattle long since dead, their bones bleached like dead coral. Occasionally, circling vultures alerted us to a black cloud of flies swarming over the fresh body of an animal that could wait no longer for the coming rain and grass.

Nothing had prepared me for the stark churning of life and death in the outback. This was no pastoral scene, but a devastating example of the biosphere at work.

The insect life was like something out of a horror movie. Flies were ubiquitous. We humans became adept at twitching isolated face muscles to temporarily remove a fly that was trying to get at the moisture in eyes and other orifices. Horses and cattle lined up in single file with the head of one hidden in the tail of the next. As the wet season approached, the hordes of flies became so overwhelming that we made ourselves fly nets to hang over our faces, similar to the ones that protected the horses. The flies would eat the skin at the corner of their eyes right off their faces if left unprotected. When out riding, the back of the person in front of me would appear jet black from the flies that had settled there. My back was the same.

Centimeter-long meat-eating ants were everywhere. Termites ate the tires right off cars. Centipedes were not uncommon, and woe unto you if you ever put on your boots without first shaking them out. There were praying mantises that stood almost a foot high. One lived just outside the kitchen door. It looked formidable. Grasshoppers abounded. Some of them had exquisite bright colored dots of yellow, red, blue, and green, as if an Aboriginal had painted them.

The population density and diversity of these critters was astounding. And so was the noise. I was awed by so much life squirming, hopping, and buzzing.

But sometimes, when I was so hot I felt I would faint, and the flies were flying up my nose and in my eyes, I wondered what the hell I was doing there. What did this have to do with being in Biosphere 2? Ah, self pity, one of the most debilitating human emotions. And then I would hear my mother: "Chin up, and shoulders back," and I would remember why I was there.

Steak was the mainstay of our cuisine. A huge walk-in refrigerator usually held various cuts of hanging meat. Soon after I arrived, the larder ran low, so we went on a "kill." Out we drove in the flatbed truck with a few branches of gum tree slung in the back to lay the meat on.

It was standard practice in the outback that you never ate your own, so we drove around until we found a stranger, a steer with a foreign brand that had wandered onto Quanbun from the station next door through a hole in the fence. Someone shot it and the other cattle simply stood and stared as if nothing had happened.

We butchered the steer right there and then, skinning, gutting, and heaving enormous pieces of meat into the back of the truck and covering them with eucalyptus leaves to keep the dust off as we drove back to the homestead. I became pretty handy at butchering and, along with Taber, inherited that job in the Biosphere.

The term *fresh meat* took on a whole new meaning. The meat had not only never been frozen, dried, salted, smoked, or otherwise preserved, the backstrap (filet) I cooked was so fresh it twitched on the cutting board. It was gruesome, but normal at a working ranch, and so it was in the Biosphere. One simply has to put squeamishness on hold when connection to one's food is so direct.

As the wet season approached, we began preparing for the onslaught of water. Day after day the air became more stifling, sweltering, and gray clouds began encroaching on the blue. Diana's dogs, usually hyperactive, lay on the steps motionless but for their tongues lolling as they panted. I was told this was the time of most suicides in the outback. We moved the homestead horses to higher ground. We went to town to stock up. I rode out with Diana on the rickety tractor to build berms and lay logs in the path of the expected flooding. This would help reduce the devastating effects of runoff and soil erosion in areas where the grass pastures were not yet fully recovered from desertification.

One day, towering cumulonimbus clouds hung in the sky while a war of the gods erupted inside them. Drops fell, huge freshwater tears. And they fell, and they fell, thirteen inches in one night, and then twelve inches in the next twenty-four hours. The waters rose up around the stilted buildings to where we could float down the road in inner tubes. There was no getting to town for anything. We were on our own.

Silke Schneider, or Safari, is one of those people whom everybody loves to be around. She is a warm, happy-go-lucky, adventurous soul, without a mean bone in her body. A tall woman with mousy, shoulder-length hair, she has particularly long, well-formed legs that I have always been jealous of. I have rarely seen her without a red bandanna around her neck and a belt with a big buckle that she wraps around whatever she is wearing, which is usually jeans. At nineteen she left her home in Hamburg, Germany, to join the American Circus and perform a cowgirl act.

I first met her in 1982, when I lived at the October Gallery. She arrived with John's entourage for the theater tour around Europe and the U.S. As she strode in the gallery door I noticed she wore a white flat hat with blue polka dots. I was twenty and she was twenty-two. We made friends instantly and hung around together so much—making theater costumes and giggling at just about anything—that people started calling us the Bobbsey Twins.

A year later, during the opening ceremonies of the Caravan of Dreams, I saw her arrive on her gray horse, Samir, wearing a cowboy hat, cowboy boots, and a red satin shirt with white fringe. She rode up right on cue, as two other cowboys were pacing off a staged shoot-out in Main Street in front of the Caravan. Safari had ridden over six hundred miles for the opening, all the way from Santa Fe, New Mexico, to Fort Worth, Texas, on historic cattle-driving trails.

Now, she was in the outback following her love of horses, and also to assimilate what I had come to learn from having very few modern conveniences at hand, no doctor and no vet on call day and night, and only one's own resourcefulness and that of those few people around to rely on.

As the rains began, I was titillated by the isolation, excited by being cut off from "civilization"—until we received a chilly reminder that we could die out there. So much rain fell that the horse paddock began flooding. Safari and I walked out to move the horses, up to our knees in mud and water. I was focused on putting a halter on one of them.

Out of the corner of my eye I saw Safari leaping frantically about, taking huge, bouncing strides as if she was on the moon. She had stepped on a king brown snake, one of the deadliest snakes in the area. She was saved by the dulled response of the serpent, whose lightning reactions were slowed from being submerged in the frigid water. During stock camp later that year a king brown killed a horse.

As the wet season wore on, billabongs swelled and fences washed out. One afternoon, on a rainless day after the flood had begun to subside, Diana walked into the homestead paddock with Dolly, a gray mare. The horse was limping badly. She had been caught in a downed fence and scraped all the flesh off the front of her leg. Diana tasked me with caring for the poor animal. It was a hideous wound—I could see the cannon bone between rotting muscle. Twice a day I hosed down the leg to clean the injury and wash out fly maggots, then sprayed it with purple antiseptic. Bush medicine was a skill I would need when I cared for the Biosphere 2 domestic animals without a vet. The wound finally healed and the mare went back out to pasture to breed.

I practiced bush medicine again when a dingo, Australia's native wild dog, removed a large mouthful of a calf's hindquarter. I was flabbergasted when I saw the ragged hole. Miraculously, we nursed it back to health. What a cruel place! Savage and raw.

But an extraordinary thing happens during the floods. Animals that would normally eat each other on sight climb to higher ground and perch atop fence posts, snake next to lizard, next to centipede, alongside mouse and frog, having called a temporary truce. I had misgivings about some of the other biospherian candidates, but I assumed that we would be like these animals, getting along because the situation necessitated it, although we would rather be at each other's throats.

The flood took its toll on us. Tempers flew. Cups smashed

against walls, and fists slammed through screen doors, and never about anything monumental. The outbursts were simply outlets for the frustration of being utterly unable to prevent the maiming that was thrust forcefully into our faces daily, to halt this rabid hell, this marvelous, unbridled orgy of life and death.

I should have learned that intense situations tend to bring out negative behavior that is simply not present day-to-day. Instead, I chalked the outbursts up to the quirkiness of the particular personalities. Had I understood it as a more general pattern, I would not have been so perturbed when, in Biosphere 2, we all began behaving in ways I had not foreseen.

Only looking back would I see specific lessons from the outback that would be applicable to life in Biosphere 2. Days before the start of the two-year experiment in the Biosphere, Safari and I had to replace a goat with Cricket, a doe that to that time had been a pampered show-goat.

I remembered when Diana, the station manager, brought a horse up from southern Australia that had been barn-raised. A new horse was always put out with an older animal to show it the ropes. The day the mare arrived I led her down to the homestead billabong for water, before taking her into the paddock to meet her caretaker mare.

I stood and waited for her to drink, knowing she would have to be thirsty after the long trip.

She stood stock still.

And she stood.

Eventually I got down on all fours and drank out of the billabong, whereupon she put her head down and gulped down the water. She had only ever drunk out of a water trough. I feared for the animal's life out in the wilds. But she stuck with the old-timer, survived, and became a great workhorse. Safari and I placed Cricket the goat with Sheena, the most calm and reliable doe, and she became one of the best milkers we had in the Biosphere.

The two months of wet season evaporated and we got ready for camp down by the river at the stockyards, chuck wagon and all. It was magical. On the first morning of roundup I awoke at first light in my swag (sleeping bag on a damp-proof mat) to the chill of a July winter morning. Red dust rose from the horses racing around the yards, thousands of sulfur-crested cockatoos and galahs (their rose-breasted cousins) screamed at each other and swirled around the ghost gum tree tops, each species of bird claiming its own tree crown.

The smell of bread cooking in the cast-iron Dutch oven sitting in the coals made my mouth water. We made billycan tea using an old coffee can hung by a wire handle over the fire. The cook on duty threw tea leaves into the boiling water, then removed the billycan from the fire and swung it overhead to push the leaves to the bottom via centrifugal force. The few remaining floaters were sunk by sprinkling cold water over them. Usually it worked perfectly, but when it didn't, most of us had teeth to sift out the tea leaves.

This was a far cry from English Breakfast, but I tried not to turn my nose up at the barbaric way in which we made tea at camp. I soon learned to appreciate it. I discovered that there is rarely a single right way of doing something.

After feeding the horses, we ate together around the fire and headed out to round up cattle. Generally we broke up into two groups consisting of around ten Australian Aboriginals with the few of us from the homestead, one Aboriginal leading each group. We rode out through wide-open plains with grass that I had helped seed up to the horses' girth.

The grasslands were peppered with eight-foot-high red termite mounds in the sand country, and smaller gray mounds in the clay. Sometimes I caught a glimpse of a flock of green budgerigars, visible to my untrained eye only when they alighted on a dead tree marooned in a pasture. Bosques of eucalyptus and acacia held shrieking clouds of white cockatoos and pink and gray galahs.

Occasionally, small flocks of no more than twenty rare black cock-atoos with red tails flew silently in front of our herd. Pairs of long-legged cranes performed their gyrating mating dance in the tall grass, singing their eerie whaling calls. Other birds flocked to the billabongs, now full of water and crammed with life.

Once back at the stockyards with the cattle, we separated out those head going to market, and branded the young ones, put tags in their ears and castrated the "micky bulls" (yearlings). You have to get used to a lot of dust, be ready to leap over the rails, and tol-erate the smell of burning hair and singeing flesh. A rancher's life is definitely not for the faint of heart.

Sometimes I felt I was getting separated out—but which corral would I end up in?

The Aboriginals were a wonderful mystery. I needed several days to understand pidgin. Between each other they often communi-cated with hand signals. Sometimes I would hear only a few grunts while Duncan, the head stockman, worked the air with his elegant hands, long fingers conveying complex instructions to one of his workers.

Traditional Aboriginals do not have much of a sense of material possessions, and will not live inside a house. When the white mis-sionaries came to the outback, they built the Aboriginals houses in the middle of the savannah, with streetlights over the roads. They didn't ask the people what they actually wanted. It is a surreal expe-rience to drive out of the dark into one of these overly lit settle-ments. All the windows are broken out of every house (why would you want to stop the breeze coming through the house?), and the buildings are used as storage. All the furniture is outside—a vivid lesson in appreciating different values, and diverse worldviews.

The older Aboriginals could tell you exactly what animal had left a track, even a partial print, and what plants were emu tucker (food) or held water in their roots. When rounding up the cattle, they always saw the herd long before I or any of the other nonnatives did, and I aspired to acquire their powers of observation. (When I finally laid eyes on the cattle I could have sworn they had been hiding behind the trees, their abdomens bulging out on either side of the trunk. I imagined them sucking in their stomachs so as not to be seen.)

The Aboriginals epitomize resourcefulness. If a vehicle gets a flat tire, and they have no money to fix or replace it, they simply stuff the tire with rags. It works. I saw trucks, full to overflowing with people, driving on rag-stuffed tires.

They also have a different concept of reality than the Protestant ethic I grew up with. They live in Dream Time, and believe in capabilities that scientists are still trying to prove or disprove. Once I was in Fitzroy Crossing, on my way with Diana to Derby, a four-hour ride down a single-lane highway. There was no other way to get there.

As we left the town, Diana pointed to a man crouched by the road, and explained that he was a Feather Foot, a secret messenger, and was sometimes charged with punishing wayward tribal members. He could teletransport, according to the Aboriginals. That sounded like a quaint superstition to me. But no vehicles passed us on the road to Derby, and only one truck passed us in the other direction, back to Fitzroy. As we pulled into Derby, there was the Feather Foot, standing by the highway. Had he flown there with a bush pilot? Unlikely.

One day during roundup, we broke into two groups. I was to ride with Malcolm, an Aboriginal who had never been on the station before. I was nervous because, if one gets lost out there, death can come quickly. I certainly was incapable of finding my way back to the camp after wandering across miles of savannah with no distinguishing marks. After exchanging grunts and hand waggles with Duncan, Malcolm explained that we were to meet Duncan and his group's cattle in such and such a clearing on the other side of the station.

Off we went, finding an unusually large herd that we success-fully rounded up and pushed from one clearing to another, each clearing looking pretty much like the previous clearing. Eventually, after riding for hours, we came to a stop in another clearing and waited. And waited.

We waited several hours. I was getting apoplectic, convinced we were in the wrong place, lost with no water and no way home. Sud-denly, I heard the rustling of grass and clicking of hooves on pebbles and there was Duncan with another huge herd. I had difficulty under-standing how Duncan, who knew the station well, had found this unremarkable clearing, let alone Malcolm, who had never been there before. They say that Aboriginals can communicate telepathically.

I did not expect to use telepathy in the Biosphere, or teletrans-portation, and the furniture would be inside a structure with com-pletely sealed windows with no wind wafting through. Still, I treasure my time with the Aboriginals. I found it extremely humbling. It helped me embrace other's ideas and perceptions, even if wildly divergent from my own, and accept people in any station of life.

They exploded my restricted notion of what it is to be human, and made the impossible seem possible. Yeah, I thought, maybe I can become a crewmember in what promised to be one of the most significant projects in the twentieth century.

One morning, the semis rolled up, were loaded with cattle, and drove off to market. We performed the annual station play for local Aboriginal communities to their cheers and laughter. It was time for me to leave.

Now I was heading back to the *Heraclitus* for my next challenge, to manage the ship's reconstruction for a voyage to gather genetic material from whales in the Antarctic.

In Puerto Rico I found the ship gutted and behind schedule. The main mast had to be replaced, the interior rebuilt, portions of the deck knocked out and refilled with cement, sails resewn, and finally, a thick layer of insulation glued on every interior hull surface. I pretty quickly earned the nickname "the Sergeant Major." I am sure it had nothing to do with my loud voice, adept at bellowing orders.

Roy Walford, the scientist of aging at the University of California, Los Angeles, joined us for training as a biospherian candidate for several weeks. Roy, in his late fifties and the oldest of the biospherian group by a couple of decades, had been at Olympic trials as a gymnast when a young man, and still was very fit. His bald head, Fu Manchu mustache, and pointed ears lent him a somewhat alien look, though his smile was quite charming.

He had a ribald sense of humor, and was erudite, quoting from the classics prolifically. Roy's specialty was aging, or, rather, how to delay aging, particularly through diet—the 120-year diet is the subject of several of his books.

In the sixties he ran a guerilla theater, and was friends with Julian Beck and Judith Malina of the Living Theatre. While in med school, he and his mathematician friend, Al Hibbs, went to Las Vegas to win money to buy a boat. They had worked out a statistical analysis of roulette wheels that, if they could find one slightly out of balance, would allow them to win more than lose. They sat for days watching each roulette wheel in the casino. Finally they found the one and started betting. They won.

And they kept winning.

At first, the casino was delighted. Food and booze were poured on them. Crowds thronged, but, as their winnings grew, the casino management became alarmed. The two of them made the cover of *Life Magazine* for the exploit, and got their boat.

When I met Roy at the Galactic Conference, he filled me with stories from India, a trip he had funded through a grant to discover whether meditating swamis could really lower their temperature at will. He made his way around India measuring the rectal temperature of holy men.

Roy considered Biosphere 2 one of the most exciting projects then happening on the planet. When he signed up to be a biospherian, John had told him to spend a couple of weeks on the *Heraclitus* in the Caribbean. So, Roy arrived in Puerto Rico expecting a relaxing jaunt through the Caribbean islands, only to discover that his ride was under construction, and he never left port. Nevertheless, he buckled down and sewed the sails.

After three months of refitting the *Heraclitus,* we sailed out of San Juan harbor. I had earned my gold star as a manager—we left Puerto Rico on time, with everything completed except the insulation, which we installed along the way to our destination, Savannah, Georgia. I was thrilled to be at sea again.

After a brief landfall on the island of Cacos—where I dived with eagle rays amid elk-horn coral—we sailed into a crashing storm. In the middle of the night, there was a call for all hands on deck—the yardarm (the boom holding up the sail) had broken, smashed onto the deck and the main sail was dragging overboard.

This was one of those moments when personal likes and dislikes fly away with the howling wind. We worked instinctively to protect our fellow crewmen from falling overboard, taking orders without hesitation, and heaving on ropes with strength we did not know we had. Finally, dripping with seawater and sweat, we hauled in the mainsail. In the coming days, we fixed the sail, the rigging, and the yardarm and kept right on sailing.

Once in Savannah, we prepared for the first release of captive dolphins to the wild. John Lilly had performed communication experiments with Joe and Rosie, two bottle-nosed dolphins. They had been in captivity for years, and Rick O'Berry, Flipper's trainer, was retraining them in the ways of the wild. We were to build them an enclosure in a protected spot off the coast for their final training.

The ship was on its way to the Antarctic, and Abigail Alling was the expedition chief for the voyage. A graduate from Yale as a marine biologist with focus on cetacean science, she was another of the biospherian candidates.

Gaie, as we called her, was a couple of years older than me. She was petite, a jeans and T-shirt kind of woman, though I saw her in scuba gear almost as often. She had a head of thick blond hair that fell just below her shoulders when she didn't pull it back in a careless ponytail. Her gray-blue eyes stared out of her pretty face straight into your soul. She usually cocked her head to one side when she talked about something mystical or moving.

Rumor had it that Gaie had turned down the lead role in *Jaws*. I don't know how she was offered the role or whether it is true, but she had the California girl-look. She came from the East Coast, though, and hung out at least part of the time in the quaint coastal town of Kennebunkport, Maine, where her parents were friends with the Bush family. That's about all I know—she didn't talk about her past much. None of us did. Gaie was, and remains, a mystery to me.

Now she was heading up the Institute of Ecotechnic's role in the Joe and Rosie release project, working closely with ORCA (Oceanic Research and Communication Alliance), a U.S. nonprofit devoted to all things cetacean. Gaie displayed the patience of Job as she worked through the serpentine bureaucracy of permits, made worse by the fact that no one had officially released captive dolphins before. Other cetaceans had been clandestinely released at the dead of night. Usually fishermen caught them again within weeks. Joe and Rosie's release was entirely successful. They were seen two years later with a calf, having joined the local pod of dolphins.

I had completely fallen in love with life at sea. I desperately wanted to go with Gaie to the Antarctic, and continue this life of oceanic adventure. But Margret gave me an ultimatum: I had completed my basic training, and either I got myself to Biosphere 2 within the next two weeks, or I would be eliminated from the pool of candidates.

So, on April 12, 1987, I left the treelined boulevards of Savannah, dripping with Spanish moss, and landed in cactus country, Tucson, Arizona, to begin a wholly different voyage.

Chapter 6

BUILDING
A NEW WORLD

"Good afternoon, I'm calling from Biosphere 2. I'd like to inquire about ..."

"Bio-what?"

"Biosphere, B-I-O-S-P-H-E-R-E, Biosphere 2"

"Never heard of it."

"It's a project in Arizo ..."

"And what is this biosphere thing, anyway?

" Well, er, we live in Biosphere 1, the sphere of life around the Earth, and Biosphere 2 is ..."

So almost every phone call began when I rang a possible vendor for the first time. The term *biosphere* was simply not in the vernacular.

The idea of a biosphere is ostensibly Russian, though some argue that the honor goes to Austria. It first appears in literature in a 1875 monograph about Alpine geology by Eduard Suess. He makes a single reference to the biosphere: "On the surface of continents it is possible to single out a self-contained biosphere." Apparently, he also coined the terms *atmosphere* and *lithosphere*. However, in 1926, Russian geochemist Vladimir Ivanovich Vernadsky first wrote about the Earth's biosphere as an integral dynamic system.

In the late nineteenth century, Russian scientist Konstantin Tsiolkovsky, who pioneered rocketry and space research, imagined humans building self-sustaining "space greenhouses." Shortly after the launch of Sputnik in 1957, when the U.S. and U.S.S.R. locked in a race to the Moon, Evgenii Shepelev walked into a small steel chamber in Moscow, and sealed the door behind himself. The chamber was no bigger than a bedroom closet, and not the roomy, walk-in kind; there was just enough space for him to sit on a wooden chair.

For twenty-four hours, the Russian physician remained inside, breathing the oxygen provided by a small tank of algae. The experiment worked, but barely. A few months later, he successfully ran a similar month-long experiment. It was not exactly what Tsiolkovsky had in mind, but it was a start.

In 1972 the Soviet Institute of Biophysics under the leadership of Joseph Gitelson continued this research in the frigid Siberian town of Krasnoyarsk. The most sophisticated experiment, Bios-3, recycled all the air and 95 percent of the water and produced half of the food for a crew of three for six months using nothing more than tanks of algae and beds of wheat. It was a long way down the road toward a self-sustaining space greenhouse.

Here in the U.S., NASA had begun experimenting with simple systems that relied primarily on algae for oxygen production. But in 1971 Dennis Cooke from Kent State University wrote a chapter on the ecology of space travel in Eugene Odum's groundbreaking *Fundamentals of Ecology*. Cooke called for life support systems with complete regeneration, including higher plants, for long-duration space exploration. Algae systems would not provide complete regeneration, as the human population would only have algae to eat, which, aside from being quite disgusting, would not provide adequate nutrition.

Cooke wrote, "The fact that we are not now able to engineer a completely closed ecosystem . . . is striking evidence of our ignorance of, contempt for, and lack of interest in the study of the vital balances that keep our biosphere operational. Therefore, future efforts to construct a life support system by miniaturizing the biosphere and determining the minimum ecosystem for humankind is a goal that is as important for the quality of human life on Earth as it is for the successful exploration of the planets."

Cue Biosphere 2. It promised to give us complete regeneration, using the Earth as its model.

By the mid-1980s, Westerners were used to seeing the paradigm-changing image of the Earth from space that the *Apollo 17* crew took in 1972–the blue jewel hanging in the black of space, so beautiful, without visible political boundaries. This was our home in the galaxy, and it seemed horrifyingly vulnerable, fragile.

But the notion that the life surrounding the globe formed a bio-sphere had not yet been assimilated by our culture, and *ecosystem* might as well have been a dirty word in some circles, including NASA. At SBV, we were pushing the limits of the science and technology of the time to build Biosphere 2. The fastest desktop computer was a 286, there was no Internet, and the smallest cell phones were each the size of a briefcase and mounted in the company Jeep Cherokees.

The early eighties were a time of promise, economic growth, and exploding electronics development. In 1981 the first personal computer burst onto the scene, as did the first nongovernment e-mail system, revolutionizing our project. NASA had successfully launched the space shuttle, the first reusable space vehicle until, in January 1986, the tragic *Challenger* explosion threw the program into turmoil. Although NASA had a stated goal of a mission to Mars, the stand-down after *Challenger* fueled a groundswell of support for commercial space exploration.

Space Biospheres Venture joined many other private organizations headed to space, with Biosphere 2 promising to be the prototype of a new kind of life support system. Also in 1986, scientists discovered a dangerous hole in the ozone layer that protects us from the most harmful effects of the sun's radiation. Our project rode the wave of the growing environmental movement, becoming a symbol of great hope. The opportunities and the challenges were enormous.

The overarching question was, is this possible? Can a life system that evolved on a planetary scale over billions of years be bottled? Alternatively, will something inexorably cause it to fail, and without explanation? Without geologic processes, will nutrients bottleneck? Without high-atmosphere chemistry, will toxins build up and kill everything? Within a small, closed atmosphere, will there be such a load of bacteria and spores in the air that the humans succumb to deadly lung infections?

From the human perspective, our earthly biosphere provides us with air to breathe, water to drink, food to eat, shelter for us and our families, and a way to recycle most everything, continuously, without producing dangerous poisons. Our Biosphere had to do all that within a hermetically sealed three-acre vessel.

We were taking an entirely different approach from NASA engineers. They compromised the complete recycling of everything. Crops would be grown in hydroponics, as automated and sterile as possible, in their system planned for the trip to Mars, and the initial outpost.

But we were designing Biosphere 2 for space colonization. Self-sustaining systems would be required, given the at-minimum nine-month trip back to Earth to restock. It would be a home away from home, a natural ecosystem with bacteria and all, ultimately recycling 100 percent of every aspect of the life support system, giving residents total material autonomy from Earth.

Biosphere 2's second objective was to provide a laboratory for ecological study—a world in a test tube. Following the earthly analogy, biomes, or ecological zones, would be the building blocks of our Biosphere, and form the basis for ecological study. *Conservation* was the day's environmental buzzword, but restoring a damaged or eradicated ecosystem was little understood. Building a mini-rainforest from the ground up would teach ecologists hitherto untold lessons.

The Biosphere was a big Rorschach test with everyone seeing it through his or her own lens. The chemists saw it as a major chemistry challenge, the psychologists as a human experiment, the architects as an architectural symbol. At its most frenetic, almost four hundred people from wildly divergent professions rushed about designing, building, installing, testing, maintaining, transplanting. There were engineers, ecologists, construction workers, architects, drafters, doctors, electricians, plumbers, welders, mountain climbers, botanists, agronomists, plant pathologists, accountants, photographers. Excitement buzzed, and people flocked to help. We were making history.

By the time I arrived in early 1987, Biosphere 2's basic layout had been solidified: seven biomes. The Intensive Agriculture Biome (or IAB for short, or the Agriculture) would provide food, waste treatment, and water production for eight crewmembers, who would live in the Human Habitat.

The five "wilderness biomes" would provide most of the remaining life support functions. Primary among these was air management–the consistent provision of oxygen to, and the removal of carbon dioxide from, the atmosphere. These five were modeled on tropical or subtropical biomes, where plant growth and biodiversity are the most abundant. We would have a tropical rainforest, a savannah, a desert, a marsh, and an ocean, all crammed into 3.15 acres. The crew would not only have to grow their own food, recycle their own water; they would also manage the air they breathed.

We broke ground in January 1987. Now road graders and bulldozers ran around the site, gouging huge loads of soil out of the earth and mounding it to the northwest of the Biosphere. There, a mountain was slowly growing. We would stand atop it to observe the embryonic Biosphere's development. At any time day or night I could hear reversing tractors and road graders beeping and diesel engines growling.

During groundbreaking festivities, Margret announced that the crew for the first two-year mission would be eight biospherians, mirroring the likely team size and mission duration of an initial Martian foray. (Although reminiscent of *astronaut* I was very happy not to be a *bionaut* as it sounded too close to *bio-nut,* or *bio-naught.*) Closure date, or experiment start, was set for 1989, leaving only two short years to design and build a complete, biologically based life support system for eight, with the ability to expand to ten. The Russians had been working on this for nigh on twenty years and had not accomplished a fully recycling system. But I was such a consummate ideologue that it didn't even cross my mind that we were attempting the impossible.

I remember the day I arrived so clearly. It was a warm Arizona Sunday afternoon, one of the rare times that SunSpace Ranch was quiet. While wandering through the campus I met John Allen. My heart leaped into my mouth as his huge frame filled my entire field of view. As the alpha male of our group, he demanded a high level of performance from everyone, and I bounced up to greet him, eagerly seeking his approval.

"Hey, kiddo, ready to make history?" John gave me a bear hug and walked on.

Was I ever! And I was absolutely determined to get on the team to live inside the Biosphere. For now, I was responsible for the entomology program, catching and breeding the many insects needed for the proper functioning of each biome, including the Agriculture area. In figuring out which insects to include, I was to work with a team of experts, specialists in different types of insects: bees, termites, ants, leptidopterans (butterflies and moths), pests, and water bugs. That was just for starters. We had to know what insects we needed, and we required an insectary where I could maintain them.

That insectary was my domain, and mine alone. It was a space-age looking building, made of a chunky, modular prefab system designed for Mars bases. In keeping with the space theme, not to mention U.S. Department of Agriculture regulations, airlocks led into and out of the breeding areas to keep any insects from escaping into the desert and wreaking havoc. On the day we celebrated its opening, Dr. Steve Buchmann, our pollinator consultant, brought bee larvae delicately cooked in a creamy cheese sauce. As he held out the dish, eagerly showing me this gastronomic delight, his face looked for all the world like a bee, staring down at me with enormous eyes magnified through thick glasses. He challenged me with a mischievous smile, but I recoiled. I could not bring myself to eat larvae.

I barely had time to settle into my plain but comfortable room in the red-tile-roofed, single-storied brick house, a home built years ago by a fellow Englishwoman, the Countess of Suffolk. The Suffolk House surrounded a small courtyard, where a growing community of almost thirty residents met every morning. We shared a communal kitchen, dining room, and library. The house sat perched on the edge of the Canyon del Oro. Over the coming years I would often sit on the grassy patio under the lone palm tree, gazing across the valley toward the jagged skyline of the Catalina Mountains several miles away.

Then I was off for a few weeks to train at the best insect zoos in the country, the Smithsonian Institution and the Cincinnatti Zoo, and finally to the Bishop Museum in Hawaii to learn the ins and outs of insect husbandry. I had difficulty "connecting" with the little cold-blooded creatures with hard exoskeletons. It was a far cry from handling horses and cattle. An acrid stench emanated from many of the glass and screen cages. How could such little critters make such a big stink? Nonetheless, I got on with it, reveling in the challenge of understanding where insects fit into the web of life.

At that time, a great deal was known about the taxonomy of insects, and how one species is related to another. An entomologist could tell you precisely how many hairs would be located on the hind leg of a given species. But very little had been published about what they eat, who pollinates what, and particularly how to breed or maintain them, other than a few species that were mostly unlikely candidates for the Biosphere. I mean, who wants a vast population of huge, smelly Madagascan Hissing Roaches scuttling noisily around one's biosphere, even if they are highly efficient recyclers of dead stuff and extremely easy to propagate? Yuck.

I was playing Noah, and not every animal was going to be allowed in our Ark. My favorite was a blue tarantula from South America that was so hairy, with such long legs, that it looked like a fuzzy cartoon character—it was the cuddliest looking spider I have ever seen and it tickled as it crawled along my arm. But it would be banned because, although harmless to humans, it had an insatiable appetite for insects.

I did keep a pet female tarantula in the insectary that I would amuse, or horrify, visitors with, and a praying mantis that would sit on my shoulder as I posed for the camera.

I worked closely with Dr. Scott Miller of the Bishop Museum to develop the list of insects to include. Scott started by writing job descriptions delineating what roles the insects needed to play in each specific biome. This included pollination, munching plants (which, when not excessive, is vital to the recycling of materials in the wilderness), recycling dead material, being food for other animals, and pest control.

We then went about finding bugs to fill each job in each biome. It was a tough task because of the paucity of information about insect ecology. Scott said, "A lot of these insects are going to be put in new environments with new possibilities and opportunities to eat new plants. We're not sure how they're going to adapt to this."

The insects became a serious bone of contention. The agronomists were convinced that the insects in the Wilderness biomes would run amok and kill the crops. The Wilderness biome design people were sure that the insects used for pest control in the

Agriculture area would crash their biome party and decimate beneficial insect populations in the Wilderness, thereby unbalancing the ecosystems. Some even lobbied to divide the Biosphere into two—the Agriculture entirely separate architecturally from the Wilderness, including separate atmospheres. But, John and Margret refused. Biosphere 2 would have a single, contiguous atmosphere like Biosphere 1, but an insect screen would be erected between the Agriculture and Wilderness biomes.

Steve Buchmann chose five species of bees to go in the Biosphere: two species of Arizona carpenter bee, two species of leaf-cutter bee, and bumblebees. The list did not include the honeybee, as a single hive needed more pollen and nectar than the Biosphere could provide.

Steve was concerned that the pollinators might overlook the flowers. Bees see UV light reflected off the petals in patterns, rather like runway lights showing the bees where to land and where to find the booty. The Biosphere's glass paneling would filter out most UV rays, extinguishing the runway lights. So, we placed a beehive inside a large chamber filled with common nursery plants in full, yellow bloom. To our relief and delight the bees found the flowers perfectly.

However, they also smashed into the glass. Hundreds of dead bees lay along the bottom of each wall. Curiously, the ones hatched inside the chamber learned the confines of their world, and buzzed among the heavy blossoms, filling the entire space with their swooping flight patterns while skillfully avoiding the glass. Likewise, we concluded, the bees inside Biosphere 2 would learn that the glass was a barrier, not a portal. We were nearly correct. Unfortunately for the bees, several panels of glass were included in Biosphere 2 to allow UV light to penetrate; the reptiles needed it to produce enough vitamin D. These panes of glass acted like beacons, calling some bees to the light, and their deaths.

A termite specialist with the Smithsonian Institution, Dr. Margaret

Collins, came to discuss the species that would be needed to recycle dead wood and leaf litter, aerate the soils, and recycle nutrients. Termites are critical to natural savannah ecosystems; in parts of Africa they account for half the savannah biomass. But they are also voracious—we did not know whether they would eat their way out of the Biosphere through the silicon sealant between the glass panels.

Margaret and I charged off together across the desert to find a termite colony to experiment on. As we gingerly picked our way through cactus near the main road into Sunspace Ranch, Margaret backed into a cholla, a particularly vicious cactus. There I crouched, next to the main entrance, picking cactus needles out of our prestigious consultant's backside! She was a great sport, and found the episode quite amusing. Besides, she had ghastly stories of working in tropical climes and finding insects hatching out of her leg. Unperturbed, we carried on our mission to collect a termite colony for the Termite Taste Test. We sealed the most aggressive local termites in numerous mason jars with silicon sandwiches containing irresistible termite food. The tiny white animals were forced to eat their way through the silicon to get to the yummy blotting paper. The silicon sealant proved safe from the mashing mandibles of even the most pugnacious termites.

While the road graders leveled the site and workers marked the layout with wooden pegs and pink flags, we had no working model for how to seal the huge structure of Biosphere 2. Design, development, and building were all happening simultaneously.

The Biosphere's skin would be built of spaceframe, a system developed by Bucky Fuller. It consists of metal bars that form a truss in almost any desired shape, with fantastic strength-to-weight ratio.

However, the spaceframe gave us sixty miles of seams to seal flawlessly. The structure would flex with every temperature change. Could we seal rigid glass to steel that would constantly move?

A hole the diameter of my pinky finger would be enough to

surpass the target leak rate of 1 percent per year. Our maximum allowable leak rate was 10 percent per year, or 0.027 percent per day. To put that in context, the U.S. space shuttle leaks at a rate of approximately 0.05 percent per day.

Once we sealed the Biosphere, the pressure inside would fluctuate wildly as changing temperatures caused the air to expand and contract. This demanded extraordinary structural strength, or some way to control the pressure. Bill "Freddy" Dempster, the chief engineer on SBV staff, developed the Biosphere "lung" concept, or variable volume chamber.

Freddy was an engineer's engineer. He loved nothing more than to chat for hours about the detailed workings of an air conditioning system, as he did once when some guys were attempting to fix an AC problem in one of the buildings. There wasn't an engineer, worker, or scientist who had worked with him who didn't respect him.

In his mid-forties, Freddy was a tall, slender, fit guy, with full lips and a head of wavy dark hair receding in the middle. Wearing jeans or corduroy trousers and checkered long-sleeve shirts, he of course carried the prerequisite pens and pencils in his breast pocket. In the evenings, I often saw Freddy and John playing Go, an Asian board game that has something to do with taking over territory. Freddy frequently won.

For the Biosphere's lungs, Freddy invented a round steel chamber the size of a large ice rink with a neoprene roof, which would be attached to the side of the Biosphere. The flexible ceiling would act like a bladder, rising as the pressure increased, and sagging as the temperature and pressure lowered. A metal plate in the center of the bladder would provide weight, thereby gently pushing on the air inside the Biosphere and creating a slight positive pressure. This would force air out through any leaks so outside air could not get in and "contaminate" the atmosphere inside Biosphere 2 with unrecorded molecules. A second skin would cover the lung. A fan could suck up on the metal plate, thereby modulating how much weight and how much positive pressure would be exerted on the atmosphere.

This all sounded quite simple in theory. Would it work?

To find out, in 1986 a small team built the Test Module. The chamber was four hundred times smaller in volume than the Biosphere. Freddy and his cadre of engineers needed it to test the new spaceframe and lung design and develop the sealing and leak-testing techniques.

One of my first memories of Taber at SBV is of him wedged in the spaceframe roof of the Test Module, wearing a green-and-yellow striped football shirt and billowy pants. He was covered in black butyl rubber, and was squeezing the sticky paste tediously into every seam to seal it. Once Taber had completed spreading the black goo that smelled like rubber tires, Freddy and Taber banged shut the airlock door (itself a prototype of the Biosphere 2 airlock), and tried all manner of ways to find leaks. Ironically, a fancy, high-tech, point-and-shoot infrared system did not work at all. What did work was incense. They ran smoldering sticks of incense slowly along each seam. White smoke dived into every tiny hole.

Soon both men reeked of incense, and the butyl rubber tape was a disaster. It was messy and impractical to install and did not seal well. Peter Pearce, who built the Biosphere's spaceframe structure, finally developed an ingenious system using silicon, the same material that I later tested on the termites. Double panes of glass sandwiched a shatter-resistant plastic film. The ensemble literally floated in a bed of silicon attached to the steel frames. When this system was installed in the Biosphere, the annual leak rate was approximately 8 percent, under the 10 percent per year maximum allowable rate. Biosphere 2 was the world's tightest building, with less leakage than even the space shuttle.

About once per century in Arizona, hail as big as an orange rains down. To test whether the glass would withstand this pelting, technicians blasted large ice balls from a pitching machine. The glass survived.

Meanwhile, the Biosphere was attracting the best minds, all heavyweight champions in their respective worlds: Sir Ghillean Prance, director of the New York Botanical and Kew Gardens (Rainforest); Dr. Walter Adey of the Smithsonian Institution (Ocean and Marsh); Dr. Tony Burgess, the world's expert on the Sonoran Desert with the U.S. Geological Survey (Desert); Dr. Peter Warshall, a preeminent ecologist and primate expert (Savannah); and, Carl Hodges, director of the Environmental Research Laboratory, University of Arizona (Agriculture), whose most recent claim to fame had been the design and installation of the high-tech agriculture exhibit at Disney's EPCOT Center. They were the biome design captains, each responsible for the design of a biome, working closely with a counterpart manager at SBV.

But when these great minds came together for design conferences in 1986 and 1987, the result was a modern-day Babel. Language problems ran rampant, with each discipline miscommunicating in its own jargon.

The engineers thought the ecologists flaky, while the ecologists presumed the engineers uptight. The engineers dealt in hard numbers and wanted specific requirements—what temperature did each biome have to be? How many gallons of water would each biome need? How big an area would each biome need?

The ecologists dealt in ambiguity, dynamic equilibrium, and flows through a system. They wanted to know what the environment would be like so they knew what plants would thrive, what temperature range would be provided and how much water would be available.

The engineers were sure the vines would strangle fans and sensors. The ecologists did not trust that the technology would not kill their all-important systems. Ecologists thought that the structure would leak so badly as to be nothing but a glorified greenhouse.

Engineers were concerned that the Biosphere life systems would devolve into green slime.

At one of the first meetings, confusion was such that the group passed a banana around and only the person holding the banana could speak.

Sir Ghillean Prance (whom we all called Iain) remembers the meetings with nostalgia: "Some of those biome design meetings were really exciting, because each time, one came away with the feeling that progress had been made, people had listened to each other, and they were very productive meetings."

The team eventually converged on the basic design of each biome and its support equipment:

- The Rainforest would be modeled on Venezuelan tepuis, with a sixty-foot roof to accommodate the huge trees. A thirty-foot "mountain," with a cascading waterfall that thundered into Tiger Pond, would mimic the steep volcanic rock outcroppings that typify the tepui region.
- The Savannah's analog was a South American grassland, with African gallery forest (the tree-lined area around the grasslands where predators prowl) and two Australian billabongs.
- The Desert would have to be a coastal fog desert where the ground is parched much of the year, but the cacti and succulents are acclimatized to an atmosphere thick with moisture like the air in Biosphere 2.
- The Marsh would squeeze sixty linear miles of Florida Everglades into one hundred feet. Water would flow from the freshwater Marsh with bullrushes, turtles, and mosquito fish, through three swamps, each with a different species of mangrove and increasing water salinity, until it finally reached the sea.
- The Ocean biome would have a coconut palm beach with white coral sand from the Bahamas, and a tropical coral reef.

- The Human Habitat would suggest a city, with a machine shop, gym, hospital, offices, and other necessities.
- Lastly, the Agriculture would combine the best of ancient tropical intensive farming with modern-day high techium.

Extinction was a big issue. Some supported designing for no extinction by using as many well-understood species as possible. The winning side wanted "species packing," that is, putting in more species than the scientists thought might finally survive so that if one species fails, another will thrive, finally reaching self-organized stability.

In engineering, this is called built-in redundancy. In ecology, there was no equivalent term because no one had ever designed ecosystems from the ground up.

Tony Burgess, the Desert biome design captain, says, "A lot of people are deeply disturbed by the notion of synthetic, self-organizing ecology. That is what this thing was all about, a cathedral of Gaia to test its self-organizing capacity, and for that we figured we would have far more diversity if we over-packed [with species] and we had extinction take its course, which of course was very painful to witness."

Even the dirt mattered. Different soil types formed a variety of habitats that would favor different species. There were eight such habitats in the Rainforest alone, and twenty-three different soil types throughout the Biosphere.

We carried the philosophy of species packing into the Agriculture. We would not rely on only one type of starchy crop—we grew not only white potatoes, but also sweet potatoes, and taro. Rice was not the only grain; we would also grow wheat and sorghum. We would be very thankful for this decision once we needed these crops to survive!

Peter Warshall, the Savannah biome design captain and the

person responsible for all the vertebrates in the Wilderness, advocated the idea that the biospherians needed a companion primate. The idea had originally come from author William Burroughs. John and Margret agreed. So the galago, a cat-sized nocturnal bushbaby, joined us.

Peter was also responsible for the food web—what eats what. Each animal we considered illustrated the huge gaps in ecological understanding of the time. At one biome design meeting, the Smithsonian's Walter Adey suggested the ultimate food web for the Biosphere's terrestrial biomes. He was promoting the idea of weasels, and he had the whole thing worked out with a hugely complex system of how the weasel would be supported. Peter was flabbergasted. He launched into a fifteen-minute rebuttal—how it would be impossible in such a small area, why each species in the proposed food web could not be in the Biosphere, while a grin grew across Walter's face. Peter finally stopped midsentence, realizing he'd been had. It was an ecologist's practical joke.

Below the Savannah, a cliff face towered over the Ocean. It was built of artistically modeled fake rock made of cement and pitted with planting pockets, perfect for plants that thrived in limestone soils. Rock figs and a host of other calcareous-loving plants clung precariously over the reef. Despite irrigation-system problems, and the soil repeatedly washing out of the pockets, the bare concrete was soon transformed into a living solar collector. "You could do this on any building," Tony Burgess says. "It's biological mimicry, taking plants that are already adapted to this environment and growing them on façades—living art in desolate urban canyons."

As the Wilderness biome designs emerged on floor plans and in reports, collection trips began. Tony and crew drove to Baja in his battered white truck to collect strangely twisted succulents for the Thornscrub (the ecological zone between a grassland and desert) and seeds for the Desert. Linda Leigh, a biospherian candidate and botanist who managed the terrestrial Wilderness biomes (Rainforest,

Desert, Savannah), led a team to thickly-forested mountains of Orinoco in Venezuela, returning with a large collection of plants, and malaria that left her in the hospital for months. She, Peter Warshall, and Margaret Collins, the termite expert from the Smithsonian Institution, went to French Guyana to collect grasses and termites for the Savannah.

On this trip Margaret unwittingly demonstrated a conflict that would dog the project—doing what is best for the project versus what will advance a scientist's own career. They were a bush plane ride and several days' car drive from anywhere, in the middle of the savannah grasslands. Margaret and her assistant George were looking for a colony of *Amitermes,* or another local species, *Nasutitermes,* the glue-spitting termite, which defends its colony from marauding ants by firing a stream of sticky stuff out of a tube on its head. This instantly sticks the ant to itself and its surroundings like superglue. They had been looking for days and had not found a colony of either one.

One day, Margaret had George hacking through a mound of what she thought was a colony of the glue-spitting termite. Bare-chested, he was hammering into a huge, concrete-hard mound, trying to find the queen—the termite that produces all the colony's offspring. After hours of sweaty work he found her. Margaret bent down to inspect and discovered that, while the workers they had been seeing were the glue spitters, this was a queen of the other species.

Until then, no one knew that the two species of termites lived in the same mound, sealing off sections from each other. This was a scientific discovery. Excited, Margaret carefully extracted the queen. Peter and Linda stood watching, thinking how perfect it was for Biosphere 2. Margaret exclaimed, "Wow, no one has collected an *Amitermes* queen!" She dumped it in a small but fatal vial of alcohol preservative while the others gaped. She was a taxonomist, and could not run the risk of losing having a specimen with her name on it. In the insectary I received a small colony of *Amitermes* with no queen.

We happily took advantage of a generous offer from the Missouri Botanical Gardens in St. Louis, which was revamping a major rainforest exhibit. With snow falling outside the dome, we dug up and balled the roots of mature trees and any other large green plant they would let us take, and dragged them hurriedly out of the warmth, through the freezing weather, and into the shelter of a cold semi. Most are still thriving in the Biosphere 2 Rainforest, thanks to Linda's care.

Linda had a thick mane of brown hair that waved down to the middle of her back, and a strong, square frame with little extra flab. In her late thirties, Linda was a botanist, and fit as a fiddle from years of fieldwork. As part of a study to assess the reintroduction of wolves into Olympic National Park, she had once spent three months in the forest tracking elk and deer, and counting piles of ungulate poop—an indication of how many animals lived in the park, and where. Another year, she slogged for months across the Alaska Peninsula, counting and collecting native plants to help the state decide how to manage the area.

I would often see Linda in the greenhouse that held the Rainforest plants, gently watering the tiny potted plants and each boxed tree. Or she held a hand lens to her eye, checking what bug had invaded one of her precious leafy charges.

She took her field notes in a small, orange, hard-back notebook. Her handwriting was so meticulous and precise it could have been used for architectural drawings. Watching her write—doggedly drawing out each letter—drove me crazy. I didn't have the patience to write like that, but rapidly scrawled across the page . . . illegibly.

Her writing hygiene did not transfer to her office. Her desk was neck-deep in paperwork. The floor was scattered with boxes of books, piles of papers, jars with seeds, and dried plant parts. Planting plans for the Wilderness biomes were randomly pinned to the walls.

Much of the paperwork involved permits and procedures so

federal and state laws would allow her to collect in the plants' native country and bring them into Arizona. Linda would have to quarantine plants in special greenhouses encased in fine screen, including the drainpipes, so as not to let loose in the desert any creature that might be freeloading on the plants.

Inspectors from the U.S. Department of Agriculture came frequently to inspect plants before they could be released. Once they left quarantine, Linda grew them out in large greenhouses, then transplanted them into the Biosphere. These were Linda's babies, and she nurtured each and every one of them, save the savannah grasses, grown commercially in Tucson.

In 1987, the Rainforest and Desert greenhouses were filling up with plants, and beginning to look like some exotic nursery.

There was a fundamental difference in design philosophy between the terrestrial Wilderness biomes and the Marine biomes (Ocean and Marsh). The terrestrial biomes were designed species by species, each individual plant and animal carefully chosen, collected and placed into the Biosphere in precise locations. Most organism choices were based strictly on ecological principle, though human preferences crept in.

The Marine biomes embodied a different approach. Walter Adey was concerned that many more plants and animals could be found in a square meter of marsh than in any of the terrestrial biomes. Little was known about the interactions between the organisms. To avoid having to design something so complex, we collected the Marsh and Ocean biomes as established mini-ecosystems, after Walter successfully tested this method at the Smithsonian's National Museum of Natural History.

Gaie Alling managed the Marine biomes and collected mangroves in Florida from an area that was about to be bulldozed for a parking lot. Crews dug up the mangrove trees and placed them into boxes—soil, crabs, snails, microbes and all. Synthetic tides flooded in and out of the boxes, simulating their natural environment.

The trees created quite a stir at the Arizona border. The border guard heard the driver say the trees were mangoes, which are illegal to import to Arizona. Roy Hodges (in charge of the installation of the mangroves in the Biosphere) and I were dispatched to the border to rescue the trees—and the snails that the border guard had mistaken for insects, another no-no.

I was on coral-collection detail for two weeks, not exactly a hardship tour. My old diving team of Laser, Taber, and I, along with Gaie and several others, went to the Yucatan on the east coast of Mexico to gather from reefs recently devastated by a hurricane. Although the collection area was already damaged, with large fields of dead and crushed coral, I could not help feeling I was violating the reef as I smashed at a beautiful coral head with my hammer and chisel.

We collected corals and all the volunteers that came with them—brittle stars, snails, and, as it turned out, octopi that we later tried to evict from the Biosphere Ocean because of the havoc their bottomless appetites wreaked.

In the Bahamas, I needed to catch fish without employing the stunning methods used by many commercial fish collectors. Stunning often kills the target fish and other reef organisms. Recalling my days of herding cattle in Australia, I set up large nets forming a chute, and a couple of us herded the fish toward it, finally corralling them into bags for transportation.

John kept us all fired up. He was an eloquent speaker who honed and polished the vision of Biosphere 2 until it gleamed and we could all see ourselves reflected in it. But he was extremely domineering, with a vile temper that would flare unannounced, so we walked on eggshells when he was near. At some point he began calling himself Johnny, apparently to remind himself not to be so harsh (he also took the name Johnny Dolphin as his *nom de plume*).

Although Margret was SBV's CEO and John the director of research, he was the unspoken leader among the core SBV management team, made up mostly of former IE staff. He ran a tight ship: we met briefly every morning of the workweek. We ate dinner together every night but Saturdays, where we discussed issues pertaining to the project.

These discourses often ended in a flamboyant performance by John, placing some aspect of the project in the context of world history as he saw it. His large body would sway and dance, framed by a flip chart, his stubby yet graceful fingers forming an image in the air, while he excitedly proclaimed that we were making history, achieving human destinies in our larger home, the solar system, and that we were creative, cooperative agents of the biosphere, assisting in the evolution of life, its launching into the cosmos.

He said we were helping create the noösphere, a Russian concept whereby, as the biosphere is the sphere of life, the noösphere is the sphere of intelligence, which proponents of the concept agreed had not yet fully formed around the globe. By bringing the living biosphere together with technology into a cooperative synergy managed by human intelligence, we would be creating the first noösphere, right here at Biosphere 2.

John drew mantras from George Mueller, the architect of the *Apollo* missions and the gargantuan *Saturn V* rocket. "Keep it simple"; "The better is the enemy of the good"; and the key that allowed NASA to build the *Saturn V* on schedule: "All-up testing," meaning that instead of testing parts of the rocket—which takes valuable time—the first major test of the vehicle was when it launched. We would use this same approach at Biosphere 2.

It was seductive. It was inspirational. It was intoxicating. His resolve and energy never flagged and, despite his outbursts, I came to care for him greatly.

In this swirl of activity, social life was almost nonexistent, but on the afternoon of December 2, 1987, as I passed Taber on his way to the Test Module, my life changed forever. On a whim, I asked him over for a beer. That evening we sat watching my tropical fish swim languidly in their aquarium, laughing about the future, about life inside Biosphere 2.

Until that night, there had been no romantic exchange between us, not even a glance. We had become close friends during our time together on the *Heraclitus,* and we were now thrown together in our next adventure. I had always respected his sharp intellect and I was more than a jot impressed that, at only twenty-three, two years younger than me and with no more education, he was teaching himself chemistry by night and running the design of the Biosphere's analytical laboratory by day.

He had become a little cherubic in the two years since his buff days on the ship, but he was an attractive man. Intense yet gentle green eyes gazed out from under a strong brow; his full lips parted readily to form an endearing, often mischievous, grin. He nearly always wore a gray canvas hat with both sides of the brim fastened up.

For a guy whose most vigorous exercise was twiddling knobs in the analytical lab, he was immensely strong. John, at two-hundred-plus pounds, liked to leap into Taber's arms on Saturday mornings before theater class, and lie there like a giant baby. Taber would toss him into the air and catch him several times.

That evening we giggled about how he had christened me Thunder Thighs on the *Heraclitus,* because I had gained twenty pounds and considerable glute muscles since he had seen me in Fort Worth. I teased him about how I was sure he intentionally would take hours to make his routine checks of the *Heraclitus* before he took the helm from me at three in the morning, while I was desperately dueling with sleep. We laughed until our sides ached and my next-door neighbor banged on the wall for us to shut up.

That evening I discovered a Taber I did not know: tender, vulnerable, attentive, and caring. Apparently, he had had his eye on me for some time, and that evening our friendship became fierce attraction. Thus began a relationship that has bolstered us both through hard times, and been a joy in good times, for nearly twenty years.

We kept our affair quiet. Not that anyone did not know, but we were tactful, and everyone else seemed complicit in silence. We learned to keep our professional and personal lives separate. Years later, when we ran a small aerospace company together, no one except those closest to us suspected that we were husband and wife.

Biosphere 1's media exploded with excitement. The image of the lifelike Biosphere 2 model against the Catalina Mountains appeared all over the world, looking as if the Biosphere was already built. Reporters came from around the globe. *Omni* magazine called Biosphere 2 "one of the most daring and imaginative environmental projects now underway." *CNN* filmed us, the *New York Times* wrote about us. We dressed in our biospherian red and posed for one film crew after another.

The project captured people's imagination, and seemed to touch some deep cultural nerve. A headline in *Newsweek* ran "Replicating Earth," as if we were playing God. Biblical references abounded. "Biospheres: In Heaven As It Is On Earth," wrote *Space World.* The Italian magazine *Scienze*'s headline proclaimed *"L'Arca Di Noe' Dello Spazio"* (*Noah's Ark for Space*). The English magazine *You* wrote of the second Genesis, and the landscape being "aptly Old Testament." Return to the Garden of Eden was a popular theme. *Discover Magazine* called the biospherians "Earth's First Visitors to Mars," and the English *Sunday Times Magazine* called the project "Spaceship Earth" and "Second Nature."

Biosphere 2 was awash in mystique. We were re-creating Eden, saving the Earth and fulfilling humanity's destiny to reach the stars, if we were to believe our press clippings—and most of us did. The expectations the world had—we had—of the project were impossibly high. But they were heady times. The Biosphere was growing, and money was flowing.

But even as the Biosphere slowly grew out of the ground, and biome design blasted ahead, we had not yet tested the basis of the whole experiment—the biological life support system of the Biosphere. And many said it would not work.

Chapter 7

VERTEBRATE X, Y, AND Z

A t the end of 1988, John Allen walked into the Test Module as Vertebrate X, wearing what had become his signature outfit for filmed occasions—a red shirt with "Biosphere 2" emblazoned on the pocket, slightly rumpled khaki pants, and a straw hat that he stylishly cocked to one side. John was the first person to be sealed in a complete biological life support system.

Margret had outfitted the Test Module with a small room supposed adequate for one person to live. The rest of the sealed chamber was bursting with plants. Some would produce the oxygen for John to breathe, others John would eat, and yet others made up a prototype system to recycle John's waste.

It was a bold move. Many thought the whole venture impossible. "He'll succumb to a terrible lung infection from all the microbes," some said. "There's bound to be some undetected toxin," others said.

"Research should be defined as doing something where half of the people think it is impossible, and the other half think, 'Well, maybe that will work,' " said Burt Rutan on Discovery Channel's *Black Sky: The Race for Space.* His most recent creative engineering feat is Spaceship 1, the first commercially funded manned spacecraft. "When there is ever a true breakthrough you can find a time period when the consensus was, 'Well that's nonsense.' So what that means is that a true, creative researcher has to have confidence in nonsense."

John exhibited complete confidence that what he had come to call the "science of biospherics" would work. Seventy-two hours after the door slammed shut on the Test Module with him inside the tiny world, he walked out jubilant and healthy.

We were demonstrating that the impossible could be achieved.

Soon thereafter Vertebrate Y, a.k.a. Gaie Alling, lived in the Test Module for five days. On the day she entered, over thirty press crews from around the world crowded around the airlock door to get a shot of the historic moment, and the diminutive blonde in a scarlet jumpsuit who dared seal herself up with nothing but a bunch of plants and microbes to keep her alive.

It was one of the hottest days on record, and an hour before Gaie was to walk in and close the door behind her, the power failed. We fired up a tiny diesel generator that could not possibly handle the load. The journalists were furious because they did not have enough power. Gaie and the Test Module almost fried as the generator strained to keep everything running all afternoon. That day, John, Margret, Freddy, and others decided that they needed an energy center to provide reliable power to Biosphere 2.

Several weeks later, Vertebrate Z, Linda Leigh, lived three weeks in the Test Module—another world record for the longest stay in a recycling biological life support system.

These three experiments could not show us that Biosphere 2 would work perfectly, but they did prove that the basic notion was sound. These experiments also tested and trained the biospherian candidate team as we set up and operated each one round the clock.

The media hubbub grew. We held press conferences. We were showing up in space and environmental books, in the Encyclopedia Britannica, *Scientific American, Popular Science,* and on the Discovery Channel. Robert Redford narrated a film about Biosphere 2. Alan Alda dedicated an entire program of his TV series *Scientific American Frontiers* to the project. Phil Donahue held one of his shows at the Biosphere. I could not bring myself to read the articles or watch the TV pieces. They were mostly positive, but were nearly always replete with infuriating inaccuracies and misunderstandings that, once printed, propagated throughout the press. One article claimed our objective was to build shelters for nuclear winter. That was a new one on us.

We weren't prepared for all the attention. Kathy Dyhr, a chain-smoking community organizer who had been with the project from the beginning, became the sole press person. "[In the beginning] John wanted no press coverage so I was trying to fight off local press coverage," she recalls. "In 1985, when the National Commission on Space video came out [which highlighted Biosphere 2] we had no press person, we had no nothing. I told John and Margret, 'This is going to hit like a Broadway hit.' And they actually believed that there would be no press interest."

Kathy wanted the story broken by a journalist who had enough sophistication to grasp the project's complexities and significance. She turned to Bill Broad of *The New York Times* who came to visit in 1986. After a half-hour interview with Kathy and others, Bill turned to Kathy and asked, "So why is food grown in there?" After she said it was for the people to eat, she says he threw his pencil and notepad in the air and exclaimed with an incredulous laugh, "You mean there are going to be people in there? I thought you were talking about a terrarium."

Bill's story was long, detailed, factual– a dream article. But in the long-practiced discipline of providing balance to the story, he threw in one paragraph about a man named Veysey who went to Synergia Ranch in the seventies for five weeks and subsequently published a book that included disparaging comments about the group living there. Some of those same people were now running Biosphere 2. Skeletons in the closet began rattling the door to come out.

Meanwhile the first steel plates that would cover and seal the entire underside of the Biosphere were welded–by a woman, among a roar of cheers. Once the steel liner was in place, each level area of the Biosphere was flooded. A pump forced air down pipes installed under each seam so air would find its way to any leak and bubble up through the water.

Of all things Freddy, the chief engineer, is proud of, he is proudest of the extraordinarily low leak rate. He spent hundreds of hours hunting down every pinhole. If all the leaks he did not find were in one spot in Biosphere 2's skin, I would not be able to stick my thumb through the hole.

The spaceframe structure was going up, like the Erector Set my brother enticed me to play with as a child. Large sections of wall or roof were assembled on the ground. Mountaineers strapped into harnesses would then climb aboard the approximately thirty-foot-long segments, clamp the spaceframe to a crane, and swing up into the desert air and over to where they bolted it into place. It took my breath away to watch the guys piloting spidery sections of roof eighty feet off the ground with hand signals to the crane operator.

Laser, who was responsible for quality control, charged around the site in his red Ford F-150, overseeing details of the construction, drawings in hand, pushing people to move faster. Sometimes construction leapt ahead of design and Laser and Bernd Zabel, the construction manager and a German biospherian candidate, would burst into the architects' office, demanding more blueprints.

Bernd recounts an incident early on, when the Rainforest greenhouse was being built. Trucks pulled up to pour a concrete pad at the entrance of the greenhouse. "TC [the chief architect] came over and said, 'This is not what I designed. I want you to stop right now.' I told him 'But TC, this has been scheduled for a week.' And the trucks were already pulling up and this was our first standoff. TC was standing out there shaking and yelling, 'I'll kill you, I'll kill you.' I was standing saying, 'Even if you tried, TC, you couldn't.' TC, I love him. From then on, I was ahead of TC with the drawings. I should have learned not to get the trucks, but sometimes I would have to tell TC, 'We have to pour tomorrow, give me the drawings,' and he'd say something like, 'I can't, I don't have the drawings.' So I'd say 'Let's make something up.' " The two would scribble on a yellow pad and give the "drawing" to the construction crew.

Little wonder Sally "Sierra" Silverstone, an English biospherian candidate who managed the architectural office, took to chewing pens to a nubbin.

Some complained the simultaneous design and construction process was wasteful—occasionally something had to be redone, like the Ocean's viewing window. After the walls to contain the small Ocean were already built, Margret wanted a huge window installed so visitors could see into the coral reef. The next day, the construction team chopped out large parts of one wall to install two windows a couple of inches thick, strong enough to hold the million gallons of water inside the Biosphere.

John Miller, the general contractor, found the process exhilarating, and highly productive. "The synergistic design and development process could never have been better exemplified than it was at the Biosphere, with design and building going on at once. A typical project of that scope probably would have been in the planning, design, and architectural drawing stages for several years after we had started construction. The projects I am familiar with that are government sponsored would not have been ready to close when we re-opened at the end of the first two-year experiment in 1993."

Computers would control the Biosphere's temperature, something relatively simple today, but then, there was no software for monitoring and control. We had to develop it using basic computer equipment that we would laugh at now. This was simply for temperature. Humidity, oxygen, carbon dioxide, toxins, ocean waves, and rain all needed to be monitored and controlled. The basement below all the vibrant greenery would be packed with equipment that we named the Technosphere. "Biosphere 2 was the Garden of Eden atop an aircraft carrier," Roy Walford liked to say.

Over the garden were 77,000 spaceframe struts with 6,600 panes of glass. Below, there were 640 tons of stainless-steel liner and 16,000 cubic yards of concrete structure, and almost 2,000 sensors, and all were meticulously drafted on construction drawings.

One sunny morning in March 1990, I sat at one of several tables in the Suffolk House courtyard sipping tea, awaiting the daily morning meeting of upper managers, who also sat clutching cups of tea or coffee. John strode in and announced to all that I was to replace Safari as the Agriculture manager and she was to become the manager of the insectary and domestic animal program. I had been helping Safari in the Agriculture, so I was familiar with it. Nonetheless, everyone sat in stunned silence, including me.

But I was also thrilled—this was the chance I had been waiting for, to manage a major design aspect of the Biosphere. My chances of being on the final crew soared. I hoped to make myself indispensable, but with that came growing responsibility. I was now in charge of keeping the eight biospherians fed.

I did not dare look at Safari, who was sitting a couple of tables over from me, nor dwell on how she felt about the change, for I was certain she would be upset. As I found out over ten years later, Safari was in fact relieved. She was tired of dealing with John.

I also managed the Biospheric Research and Development Center, responsible for coordinating operations. The biospherian candidates ran every aspect of it—the Test Module, greenhouses, and tissue culture and analytical labs. It was like herding cats. But whenever a tour of dignitaries came through, we all pulled together to clean the place up, put on our uniforms, and perform.

There I would stand, in what looked like a cross between a scarlet prison jumpsuit and a *Star Trek* uniform, proud to be wearing the insignia of a biospherian candidate but cringing at its theatricality, and pretend to work in the sweet-corn field. As I saw the grand poobah walking up with the escorting entourage, I'd gather up my basket of perfect sweet corn and nonchalantly wander over and launch into an explanation of the Biosphere 2

Agriculture. It was a photographer's dream, but some visitors thought we always wore our suits, which they found ridiculous.

One tour brought a group of NASA officials. They seemed impressed with what they had seen. However, for fear of giving away some ill-defined trade secret, taking photographs was forbidden, and the Test Module computer screens displaying any data were covered over with black cloth. It was embarrassing. Showing a carbon dioxide graph would not have demonstrated anything other than the system worked. This blatant snub sewed the seeds of mistrust at NASA.

Mistrust did not stop with NASA.

People who came to the project seemed to smell something in the air, something they could not put their fingers on. Secrecy pervaded the project, and especially the core management team. They were mostly people who had come to SBV from other IE projects, each with an extreme case of idealism.

What we did not openly talk about was that we were not only saving the world by reversing desertification in the Australian savannah, or building the historic Biosphere 2 project. We were creating a new way of life, a new civilization based on the notion of social synergism.

They were Synergists. I was a Synergist. It was the ultimate in egalitarianism. Every one of us contributed what we could to any project, following Bucky Fuller's concept of synergy. The result would usually be greater than the sum of its parts. It embodied William Burroughs's and Brion Gysin's *Third Mind,* the coming together of two minds to create a third, more powerful one that would generate concepts and visions that each mind was incapable of on its own. It was a powerful tool, used to great advantage at Biosphere 2.

Before "teamwork" was the rallying cry of management consultants, we held annual Synergist conferences to learn and practice ideas of working in what we called creative groups. "Leave your ego at the door," was our mantra, and we learned to shift roles abruptly to make the team more effective. The ideas were from Wilfred Bion, a psychologist who studied small-group dynamics in military hospitals and elsewhere during the 1940s and 1950s. He observed a phenomenon he called "group animal." We are all familiar with the idea of "group think." This is "group do."

We broke into teams, each to complete a specific task in a short amount of time. Sometimes, extraordinary feats were accomplished. Often, the most laughingly absurd results transpired, which we would then dissect in gory detail, trying to analyze and recognize patterns within the mal-aligned group behavior.

Once at a gathering at Les Maronniers in the South of France, a group of four full-grown men was given a task to find something that would increase the efficiency of a woodlot. After several hours of hard work, we all gathered around in a small clearing in the wood as, with earnest and serious expressions, they demonstrated their highly sophisticated piece of equipment . . . that could chop spindly twigs into kindling. We guffawed.

"Good managers are in short supply." John would often say. He pushed anyone with an ounce of ability forward as hard as the person could stomach. Margret had been his protégé for years and had been a good student.

John pushed me, right from when I first met him all those years earlier at the October Gallery in London. I absorbed it all, working fiendishly to understand and put into practice the ideas I heard. The confining schedule seemed natural to me, having lived in boarding schools for so long.

However, I dreaded Sunday evening speeches. More and more, it seemed that the speeches were designed to keep John happy, so they were generally vapid and gung ho. Sometimes he became enraged at something someone said and he would charge out of the room, slamming the door, yelling that we were all a bunch of negative shits. You knew that there would either be a big scene the next day, with him ranting, or we would be up late that night having a "group discussion" led by some top manager, about the fallacy of the statement and why we were all so negative.

John's outbursts were few and far between early in the project, but increased in frequency and severity as the pressure built. I did not allow myself to give it much thought. I was too busy trying to stay on John's good side, as ultimately, he and Margret would chose who would go into the Biosphere. That was the Sword of Damocles that hung over all our heads.

We did not flaunt our unusual lifestyle, I suppose out of fear of what the rest of the world might think. "The secret protects itself," was one of John's favorite Sufi sayings. But the secrecy was drawing attention to our secret.

Rumors abounded about what went on in the old bunkhouse where all thirty or so of us Synergists would converge on Tuesday and Thursday mornings to meditate, leaving rows of shoes outside the front door. Later, under orders from Margret, we arrived at the rear of the building, leaving our shoes on the back porch.

One morning a pig escaped its neighboring pen and either chewed or ran off with most of the shoes. We took our shoes inside after that. I suppose we felt that hiding our shoes would halt further rumors, although thirty men and women were still arriving at the same place at the same time in the early morning.

Although in 1985 there was a slew of bad press in Fort Worth claiming that the group that built the Caravan of Dreams was a cult, most people had not yet connected us with that gang. In 1986, a brief article in a Tucson paper suggested the unusual group running the Biosphere was really a wandering theater troupe, and that our scientific endeavors were simply another artistic expression.

Thankfully, the rest of the media and scientific community did not take the bait. Most put the eccentricities down to the simple fact that big projects attract colorful characters, and some even thought it lent the project mystique.

The vision was so strong, the intent so palpable, that it seemed that everyone wanted to take a peek or be involved. Gerry Soffen, who had headed up NASA's Viking Mission to Mars, said that it felt like being at NASA in the *Apollo* years. Queen Paola of Belgium honored us with a visit, as did the Grand Duke of Luxembourg. American astronauts and celebrities dropped by. Eminent scientists wanted to participate. We stood in our red suits and performed for them all.

The attention spurred us on. So did Margret. "She was good. She was tough," says John Miller, the general contractor. "She was always looking out for the best interest of the project, to accomplish the goals of the mission. I thought her intentions were always honorable. She was hard. Sometimes she would give us impossible deadlines and my mistake was that we met them a few times." Margret also knew how to throw parties to celebrate accomplishments and thank hard work—almost one hundred during the seven years of construction.

Taber worked day and night designing the world's first paperless, fully integrated, nonpolluting analytical lab. The reagents used in analytical labs are generally toxic, and pressurized cylinders deliver the gases used to run the equipment, such as oxygen and helium. Taber had to generate the gases in the Biosphere. As any toxin put into the Biosphere would find its way into our air, food, or water, he had to find nonpolluting ways to perform tests. His objective was to analyze the constituents of the Biosphere before and during the two-year experiment. This included testing materials used inside the Biosphere from carpets to cups, to make sure they did not give off any toxic compounds.

There was much discussion about whether the lab should be inside the Biosphere, or built and operated outside. Taber made and won the argument that having the analytical lab outside flew in the face of the project's philosophy.

I wanted to make sure the crew would not end up at the end of two years eating off banana leaves after all the plates were broken, so I walked to the top of the spiral staircase by my office, and chucked the supposedly unbreakable crockery onto the concrete floor below. All of it shattered but one set, which became the black and white dishes we used in Biosphere 2.

I was working with the researchers at the University of Arizona's Environmental Research Laboratory, or ERL, to complete the Agriculture design before it was time to lay it out and start planting. Just half an acre would feed us a complete diet, including all amino acids, minerals and vitamins, for every one of the 731 days (one year would be a leap year) that people would live inside the Biosphere.

The Agriculture design had required the ERL scientists to turn their thinking on its head. Ed Glenn, who ran the ERL side of the Agriculture design from around 1989, recalls, "The trials we were doing were yield trials, but they weren't saying, 'Where do we get our fertilizer? How can we recycle our waste materials? How can we keep the soil fertile?' So I brought in one of John Jeavons' guys from Ecology Action—he had a concept of growing a complete diet on a minimum area, so I hired one of his master gardeners to show us how to do it. He was here for four or five days at ERL. His job was to stand there in the sun, doing the double dig, and everybody would come to him and ask questions."

The belated realization that the project had to farm organically led to a rude awakening. Neal Hicks, the head of the ERL workers at the Biosphere, says, "Somebody [at ERL] had done the numbers, looking at what the maximum yield of corn that had been attained, the maximum yield of beans and other crops, and out of that, and with some preliminary calculations, the half-acre was determined to be enough, conservatively, to feed eight people. One day I realized, 'Oh, we can't use hybrid seeds—the most productive crops are often hybrid—or pesticides, or fertilizers.' Our half-acre projections were based on using hybrid seeds, and fertilizers, and pesticides. . . . All of a sudden that half-acre looked a lot smaller. By then we were so far along, and committed to the half acre, we had to try to make it work."

Finding equipment on the World Wide Web is so quick today. It does not take too much hunting and pecking to find the most specialized instrumentation. Fifteen years ago, I worked the phones, calling from one vendor to another, explaining what a biosphere is, and trying to find a rice dehuller that would not need a ton of seed to function, but would do more than a hundred grams an hour. I needed threshers, seed sorters, oil presses, drying ovens, flour mills, harvesting equipment, and all of it of an unusual size—farm equipment is mostly made on a laboratory or commercial scale and not much is available in between.

I finally hit on equipment made to ship abroad for villages in developing countries. It was a big day when we got our first "facsimile machine." It made communicating with manufacturers much more effective.

In 1989, on the twentieth anniversary of the U.S. landing on the Moon, conveyor belts poured the first soil into the Biosphere. As the earth rained into the four-foot-deep Agriculture soil bed, a concrete contractor stood next to Neal Hicks, watching. Neal recalls, "I thought he was probably saying to himself, 'Oh, that's eight yards a minute going in there.' But suddenly he said, 'Gosh, there is the first living organism going into the Biosphere.' He was not thinking in terms of the mechanics of getting the soil in, rather the symbolism of the first soil going in." The soil, as John liked to remind us, would be the largest living thing inside Biosphere 2.

Most of us think of soil—dirt—as the ugly stepdaughter of our natural world, if we think of it at all. However, soil literally does much of the dirty work in our ecosystems. It's the waste-recycling depot; it's the foundation for productive agriculture. In one teaspoon of grassland soil live five billion bacteria and twenty million fungi. The soil gave Biosphere 2 its distinctive fragrance, reminding me of the smell of rain on a hot desert day.

These seven years of design and construction were the golden days and months at the Biosphere. There were parties, celebrations, visitors, accolades, heated discussions, backslapping and high fives. Everyone loved us; we could do no wrong. We were behind schedule, but we were doing it. We were developing a ground-breaking ecological tool, and paving the way to Mars. "[The] Biosphere 2 enterprise is the most exciting scientific project to be undertaken in the U.S. since President Kennedy launched us towards the Moon," cooed *Discovery Magazine*.

Then Marc Cooper dropped a bombshell on us all.

Chapter 8

THE TALLEST NAIL
GETS HAMMERED DOWN

"TAKE THIS TERRARIUM AND SHOVE IT," shouted the headline Marc Cooper crashed into our world on April 2, 1991. This *Village Voice* article reeked. It stank. It was naïve. It was prudish. It was an outrage.

Cooper turned everything on its head. He called the team building Biosphere 2 a cult, John Allen its guru, and the project a stunt and unscientific. He lambasted the science community for what he considered moral corruption because Ed Bass was paying them to help design the thing.

Other media ate it up like a pride of lions thrown a hunk of bloody meat. They let rip and let us have it. It was like an impassioned love affair gone awry. All those failings they had overlooked in the fire of lust became larger than life.

What once had been the very magnet that attracted starry eyes was derided. Instead of being a forward-thinking group forging a way to the stars for the betterment of humanity, our project became an elitist cult trying to escape nuclear winter by building a self-sustaining ashram on Mars for themselves.

At Biosphere 2, Marc Cooper's name became synonymous with the devil himself. His very name made nerves jangle as if a hundred fingernails were scraping down a screeching blackboard. I tried to ignore the pandemonium. In fact, the more the rest of the world said we could or should not do it, the more determined we became.

Look at what we had accomplished already.

The biospherian team had been announced six months earlier in November, 1990. CNN broadcast the gleaming faces of the first four men and four women who were to be sealed inside Biosphere 2 on September 26, 1991. Bernd Zabel was to be the captain, Sally "Sierra" Silverstone the cocaptain. With them, in the brilliant red suits designed by William Travilla, the creator of Marilyn Monroe's famous twirling white skirt, stood Mark "Laser" Van Thillo, Roy Walford, Taber MacCallum, Linda Leigh, Gaie Alling, and, yes, me, Jane "Harlequin" Poynter.

The only thing marring that event was a last-minute crew change that left some of us stunned, and Roy deeply depressed. Press releases had been distributed and a front-page article printed in a German newspaper showing the smiling face of Safari, the biospherian from Hamburg. Nonetheless, the day before the CNN announcement, Margret yanked her from the crew with no fore-warning and no good explanation, and replaced her with Gaie. Safari was devastated, but as she later told me, "During interviews before this I was often asked, 'What would you do if you are not chosen to go in?' And I always said, 'Well, I am here for the project. I would be very upset, but I would still stay through the two year closure.' So that is what I decided to do."

Hundreds of people had worked tirelessly to get us this far, often-times way beyond the call of duty. One person had given his life, falling from the Agriculture spaceframe after forgetting to clip his climbing harness to the steel struts. I watched him fall, bouncing down through the spaceframe. He did not cry out. Today, a memorial com-memorates his life just outside the structure where he met his death.

The Energy Center was up and running, providing hot and cold water for thermal control and our electricity. Solar power had been part of the initial designs but was abandoned, as the five to six mil-lion kilowatt-hours a year needed by Biosphere 2 would have required acres of panels. Instead we used advanced natural gas or diesel generators.

All the Wilderness area was now fully glazed and leak-tested. We had begun growing food in the Agriculture and had survived the winter from hell when farm crops had been flattened by snow because the roof was still open. Low-temperature alarms sounded almost nightly because the propane heaters in the orchard did not keep the plants warm enough. Wind and cold bit through the plastic-covered spaceframe. But the plants all made it, and the Agriculture and Habitat were now almost finished.

The award-winning vaulted-roof building that had been the main construction management center was now being officially called Mission Control, following NASA tradition. Mission Control was the central command center for the crewed missions and housed the nerve center where the myriad computers received and stored all the data from Biosphere 2. It was up and humming, providing the low-temperature alarms that blared over our two-way radios.

Most of the corals and fish were in the Ocean, and seemed to be settling in well. Night after night, divers had battled the volunteer octopi that were eating everything else. Incredibly smart animals, they did not want to be caught, and never were. We eventually stopped seeing them. All the terrestrial biomes had been planted and the Marsh was full of mangroves. A volunteer curved-bill thrasher, a local songbird, chirped away in the Wilderness.

The Biosphere was awakening.

And now, here we were, receiving a powerful left hook from out of left field. I am extremely uncomfortable with the idea that we were a cult. When I think of a cult, the word calls up a brainwashed group of people engaging in what society considers bizarre and dangerous behavior—in short, a bunch of religious wackos living an alternate reality. I recall the awful tragedy at Jonestown in 1978, when over nine hundred people committed mass suicide, because Jim Jones said so. I think of the Heaven's Gate members who killed themselves in 1997 because their leader, Marshall Applewhite, promised that a UFO hiding behind the Hale-Bopp comet would carry their liberated souls to another planet where they would reincarnate. Or the Raelien sect who, believing that humans were created by aliens, recently claimed to have created human clones.

Clearly, the Biosphere inner circle had little in common with these groups. The people who came to the IE projects, including Biosphere 2, were all attracted for different reasons, but almost all had the common thread of wanting to get out in the world and do something useful, meaningful.

Throughout the history of IE, the Synergists bucked the trends, often running counter to the counterrevolution. There were never more than two hundred Synergists involved at any one time in all the IE projects. Having small numbers of people thinking they were saving the world where others had failed, living in remote locations scattered around the globe, fed an isolationist and elitist attitude.

And some people were truly eccentric, including John. Although there were no love beads and no hippie flower power, names like Captain Fun, Gitana, Windy, Golden Brown, and Firefly added to the image of people who had never quite left the sixties. Often sporting a fedora hat, John seemed to have modeled himself on the Beatniks.

October Gallery in London was the first project where the Synergists found themselves under the scrutiny of the general public. But it was an art gallery, and the Caravan of Dreams a performing arts center, and artists are allowed, perhaps expected, to be eccentric.

Biosphere 2 was the first IE project to be involved in a major way with mainstream science, business, media, and the general public. Clearly the members of IE, the Synergists, had gone a long way to transforming themselves into people who could run a high-profile project. But the isolationist attitude lingered. Remaining echoes of the counterculture mentality bred a sense of superiority that seeped out only to be soaked up by people such as Marc Cooper.

What confused people all the more was that Biosphere 2's magic—

and possibly its Achilles' heel—was that it was not conceived as any single thing, making it impossible to pigeonhole. It was a scientific project, a tool for furthering our knowledge of ecosystems and systems ecology. It was an artistic expression in its extraordinary architecture. It was business enterprise, meant to make money from spin-off technologies and later, tourism. It was an educational tool to inspire people of all ages. And it was an engineering project, developing a prototype for long-duration, self-sustaining space bases. If you ask twenty people who were part of the project what the aim of it was, you would receive close to twenty different responses.

So, the question remains, were we a cult?

The real difficulty in honestly answering the question lies in the definition of *cult*. The meaning is so diffuse that it is nearly useless. However, the predominant flavor of the word is pejorative, which I wholeheartedly reject. Those who study cults today make a clear distinction between dangerous cults and other forms of tight-knit groups that can include corporations.

Some of the common denominators between definitions of cults did fit our group. There is usually a domineering charismatic leader, a sense of isolationism, and a central ideal. John had been our unquestioned leader and was increasingly authoritarian. Before coming to Biosphere 2, I had seen John only a few times each year on his rounds through each IE project. He could be mean and humiliating, but he was also funny and inspiring. But now John remained at Biosphere 2 most of the time. His grip on the group tightened with every piece of bad news.

The isolationist attitude was particularly acute toward people who questioned our way of life. Our central ideal was the way of life itself. But I can say unequivocally that we were not a cult if the definition includes brainwashing and loss of individuality. And we certainly were not a cult based on G. I. Gurdjieff—an early-twentieth-century Armenian mystic with followers in Europe and America—as some claimed who heard that we read some of his works.

Today we have available a new term, *intentional community*, that fits anything from communes where monetary resources are pooled, to cooperative housing projects springing up all over the country. It is such a respectable term. We Synergists lay somewhere in the middle, between the two extremes of today's communities (we did not pool money).

There is a Fellowship of Intentional Communities (FIC), which listed over five hundred groups in the U.S. in 1995, some having started as early as the 1930s, and the list is growing rapidly. One of FIC's publications attempts to dispel myths, including the idea that intentional communities are cults. The publication states that:

"Although the term *cult* is usually intended to identify a group in which abuse occurs, its use frequently says more about the observer than the observed. It would generally be more accurate if the observer said 'a group with values and customs different from mine; a group that makes me feel uncomfortable or afraid.' " Although this sounds self-serving and defensive, there is a good deal of truth to it. It is too bad that most people are not able to recognize or admit to feeling threatened.

But in the late 1980s and early 1990s this vocabulary was not available. Most people had no other frame of reference than the communes of the 1960s and brainwashing cults they feared would steal their kids. We appeared radical—no one had seen an intentional community attempting a world-class project.

Tony Burgess says, "The dictionary term is 'a group that is led by a charismatic leader.' So sometimes you were a cult, and sometimes you weren't. There were times when John Allen was struggling for control, and everybody was looking for him to tell them what to do, and to cue the performances as the director of the theater, and that was cult-like. There were other times when you guys were obviously doing different things and people were operating quasi-independently, trying to run cattle stations, and that was not cult-like at all. That was a community. The interesting thing about your group, was that depending on the time and place and personalities involved, you knew how to shift modes from cult to community."

Others who were involved with the group from the early days say unequivocally that we were a cult. Biospherian Roy Walford was convinced that it was one, but, as he liked to quip, it was "the cult that built Biosphere 2."

Neal Hicks confirms that many at the project called us a cult, though he personally was not threatened by it. "There was not an individual person that I wouldn't have welcomed into my home, or wanted to go on a trip with. They were courteous, they were warm, they were intelligent. I was never exposed to any inappropriate behavior by anyone. I did not understand The Group. As a group there was a boundary put around them that you could not break through, but you could break in and become friends with an individual. So the term *cult* was used by the press, myself, and everybody else at ERL or anyplace you'd go. I had that conversation with a lot of people. And I would always say, 'As individual people, there is not a throwaway in the bunch. I like them all.' "

Others, who did not want to be named for fear of reprisal, concluded that while in the early days it was a "collegial association of like-minded people working voluntarily together in projects," it was spiraling into a personality cult. Bernd Zabel says, "It was a work democracy that turned into a seventeenth-century court."

When asked what got him hooked and why he stayed with the group so long, TC says, "One night I'm washing dishes and I said to myself, 'Wow, this guy John sure has incredible visions, and I don't have any visions, so I'm going to go along with his until I find some of my own, or don't.' So that's what I did for the next eighteen years. It was a magical experience. Hardly a day goes by that I don't use something that I learned with that group of mad people with John Allen."

At the time Cooper's article came out, almost every single one of us strongly denied that we were in a cult, and were deeply offended if anyone suggested it. None of us would deign to be part of something so demeaning as a mindless cult. Even calling us a "group" was taken as an insult.

I think I was more concerned about what other people might think if they thought that I was a cult member, than I was about the notion itself. If I was a dupe of this man who was using me as a pawn in his grand vision, so what? It was voluntary, and I was involved in what many considered the most exciting project on planet Earth at the time. And up to this point, John had been the spark plug keeping this whole crazy engine fired up and moving forward.

The truth is that John displayed extraordinary skills of manipulation, and he scared many outsiders. Tony Burgess was one of them. "John Allen knew how to play us all. He is one of the most skilled mastermind manipulators I have ever seen. . . . That's dangerous, because that is ultimate power. Power to control people's meaning basically controls everything."

Even before the blatant accusations, the Synergia was the skeleton in the closet that people could sense, however dimly, when they came to the project. The fear that our philosophy nights, our morning meditation, our acting classes might come to public light and call the whole project into question was apparently so strong that it became a self- fulfilling prophecy.

While Cooper's diatribe was a terrible shock to some of us, the public relations people around us knew we had been living on borrowed time. They knew an attack on the Synergists was inevitable; the question was how we were going to respond.

Our reaction fueled the flame sparked by the Cooper article.

Kathy Dyhr had a shrieking battle with Margret. "I put the proposition to Margret very directly," Kathy said. "Either honestly say what you are doing, or stop doing it." This was not the first time Kathy told Margret as much.

Margret did not listen. Instead, Margret told Kathy to call us "a collegial lifestyle," not a commune. Kathy says it was a mistake:

"Since we denied being a commune, some folks imagined we must be hiding something worse than a commune. This would come back to bite us later."

John and Margret attempted to diffuse the cultish appearance of this odd community by splitting up our Tuesday morning meditation sessions to two locations instead of the usual gathering at the old bunkhouse, telling us that it was entirely voluntary.

But when, predictably, fewer and fewer people began showing up, John threw a tantrum, accusing everyone of being negative, bellowing like an angry bull, "You have no discipline, no interest in the Synergia." Meditation resumed as normal.

Weird rumors swirled around us. Apparently, we were so paranoid that we had built the ridge along the northwest of the Biosphere to stop passersby on the nearby highway from spying on our nefarious activities. It had really been built at the behest of county building inspectors, who indeed wanted the Biosphere hidden from view, because the building code demanded it. Freddy would drive out to the highway and radio back to John Miller, "Well, I can still see it from the highway," so Miller kept adding dirt to the mountain until it was no longer visible.

Some reporters hurled accusations that we were unscientific. Apparently, because many of the SBV managers were not themselves degreed scientists, this called into question the entire validity of the project, even though some of the world's best scientists were working vigorously on the project's design and operation.

The critique was not fair. Since leaving Biosphere 2, I have run a small business for ten years that has sent experiments on the shuttle and to the space station, and is designing life support systems for the replacement shuttle and future moon base. I do not have a degree, not even an MBA from Harvard, as John had. I hire scientists and top engineers. Our company's credibility is not called into question because of my credentials; we are judged on the quality of our work.

As Tom Lovejoy, a dominant figure at the Smithsonian Institution who was to become chairman of the project's Science Advisory Committee, said to Marc Cooper, "What they did before doesn't really bother me. As long as there is something orderly and valuable in what they are doing now."

But this seemed to be a minority point of view. It seemed that people had forgotten that this was an entirely privately funded organization. Some of us felt that it was nobody's business how Ed spent his money.

The accusations were especially upsetting when the media paraded opinionated scientists who knew little about the project. Such articles were usually replete with misunderstandings and misinformation. Even though we all know that we should not believe everything we read in the newspapers, in reality, people do. Once a piece of misinformation was out there in the world, it seemed to propagate, showing up all over the globe, even to this day.

Clearly, the PR could have been a good deal smarter. John was a terrible spokesperson for the project, being practically incapable of making a statement that was not grandiose or smattered with historic references and symbolism, which were all too easy to misinterpret. He would not stay on message. Kathy Dyhr says, "You never could tell what would come out of his mouth. He might talk about Mars colonies, go off on ancient civilizations, go off on some riff about how terrible American civilization was. They were totally off the subject."

Eventually, John himself became an issue. NBC ran accusations from a former Synergist named Cathleen Burke that he had become a "power-possessing, violent tyrant," who dominated his followers with physical force. The only reply from John? Footage of him on NBC turning his back and walking away from the reporter.

When John, Margret, Ed, et al. started the endeavor, none of them had any notion of how popular the project would become.

There had been no provision for handling the media, and tourism was not even considered on the business plan. Now they were in the unenviable position of having their unusual management team and extraordinary project viewed through a highly critical mainstream lens. The very way of life that had given birth to the idea of Biosphere 2 was threatening to undermine its credibility.

It felt to us like bullies stomping on our sandcastle simply because it was the best on the beach. It made me furious then, and it makes me furious still. The Japanese have a saying: "The tallest nail gets hammered down." We had stuck out, becoming the darlings of all and sundry—it was only natural for us to get hammered down. With only five months to go before the Biosphere was to be sealed, the mood at the project was dark and defiant. But most of my cohorts and I put our heads down, shoulder to the wheel, and got on about finishing what we had begun.

Chapter 9

CLOSURE

I n late 1990, the biospherian team started a series of Reality Experiments, simulations that ran for a couple of days to a week, focusing on some aspect of the Biosphere and its operation. The simulations were intended to let us feel what it would be like to live inside the Biosphere.

We had already completed Eating Reality, where each biospherian took turns cooking a full day's meals using only Biosphere 2 ingredients. It was a shock to the system.

Most of us got nauseating caffeine-withdrawal headaches. We were cranky from start to finish, as our high fat, highly carnivorous American diet was suddenly replaced with an extremely healthy high-fiber, primarily vegetarian diet. We monitored the daily intake of calories, amino acids, vitamins, and minerals using a computer program that Roy Walford had developed for his life-extension diet. It confirmed that our Biosphere 2 diet was complete in all things but B_{12} and D, which we supplemented.

Nonetheless, stomachs rumbled and everyone grumbled. The Agriculture had not yet been fully glazed, and rain poured down onto the crops and into the basement where we sat on chilly January days, huddled around a makeshift setup for meetings and monitoring the temperature in the Biosphere. We could not move to a dryer place because the rest of the Biosphere was overrun with people and our offices were not finished. A French camera crew ran after us filming our every move, including our grumpy, gray, miserable faces as we hunched in the rain.

Morale was low. It was unclear what we were learning from these early operational trials, which left key project managers working way under par for weeks at a time. We whined to each other about how unwise it was to pile Reality Experiments on top of completing designs and troubleshooting construction issues, on top of performing plays and endless evening discussions.

On March 14, 1991, TC left Biosphere 2. Of course, many of the original group had gone on to other things, but TC had seemed like a lifer, a golden boy who had been at the center of the action, and he had been a crucial part of the project.

TC and Margret, who considered herself coarchitect of Biosphere 2, had known each other for almost twenty years. "[Margret and I] had daily, daily battles. My conflicts with Margret were so great that I could not stay there," recalls TC.

He had already almost left in March 1989, over an incident when his girlfriend, Michelle, came to the project. "I had arranged to go pick up Michelle at the airport and Margret changed the deal and told Sally Silverstone and Safari to leave without me. Just the whole idea of someone meddling around in your private life like that! I came very close to leaving. I walked out across the desert with my backpack on, and stashed a bunch of stuff that I was going to come back and get later.

"But I felt it was important for me to stay there. Without my license we could not build it, unless you could find someone else nutty enough to turn their license over to the project, which was not likely. . . . A year later we had another blowup, and I realized I was going to leave as soon as I could. I thought I was dying. I was physically falling apart. It was so unpleasant."

It could only have been more shocking if Margret or John had left. Did this signal the beginning of the end of the Synergists concentrated at Biosphere 2, or simply the culling of some dead wood that had outlived its usefulness? I had no idea.

None but those in the inner circle felt the infighting—everyone was too busy. It was thrilling to be in the middle of all the excitement. I would walk into the Agriculture with my hard hat on, ready to plant a field of potatoes, and marvel at the Biosphere continuing to grow up all around me. I stood in a crowd of several hundred people, clutching a glass of champagne and cheering as the last piece of spaceframe was placed on the Habitat library along with the American flag. The volunteer cats and some sparrows had been caught and released into Biosphere 1.

The Agriculture was finally glazed and leak-tested and the soil-bed reactor was turned on for testing. The entire Agriculture soil bed was designed to allow air to be forced through the soil from below. The microbes in the earth would take water-soluble pollutants out of the air and break them down into nontoxic compounds to be assimilated harmlessly into the life cycle of the Biosphere—a biological means of cleaning the air. Sometimes I would stand gazing across the Agriculture imagining all those billions of microbes at work.

As the Agriculture was now essentially sealed, the atmospheric effects of the soil-bed reactor could be tracked. Much to everyone's alarm, the carbon dioxide (CO_2) level in the Agriculture was rising. It had risen from the ambient level of 340 parts per million, or ppm, (as I write this, the Earth's ambient CO_2 has risen to over 370 ppm) to over 2,000 ppm and was not showing any signs of coming down again. To account for all sources of CO_2 aside from the soil-bed reactor, everyone who worked in the Agriculture logged their time in and out. I tracked all soil disturbances, such as turning a plot in preparation for planting.

Every square inch was planted with something green and growing to absorb CO_2. It kept rising.

On April 30, 1991 the CO_2 rose above the yellow alert of 6,000 ppm to 6,093 ppm. This is still safe for people to breathe.

Red alert was set for 8,000 ppm, when sensitive people can get headaches. We were all sweating bullets. A CO_2 problem of such magnitude would end the two-year experiment before it even started.

On May 3, the CO_2 reached 7,400 ppm. Laser turned off the soil-bed reactor at one in the afternoon. The CO_2 started dropping. We loosed a collective sigh of relief and mopped our brows.

Apparently the problem was only associated with turning on the soil-bed reactor. If we prevented the air from flowing through the soil we should be able to manage the CO_2 in our atmosphere, but it was becoming clear that our atmosphere would need active management.

Despite the tense and sometimes belittling mood, the biospherian team was building a strong rapport and camaraderie. We wrote a play, entitled (of course) *The Wrong Stuff*, performed to gales of laughter from many of those working at the project. It was a spoof, taking everything we thought could possibly go wrong and enacting it in outrageous costumes and silly songs.

The climax came when the crew's captain, portrayed by Bernd Zabel, our real-life captain, leaped off the cliff to his demise, and the rest of the crew mutinied against Mission Control and declared independence from the outside world. "Cut the cable, cut the crap!" we chanted over and over again.

As Bernd recalls, "Theater was originally a way to deal with emotional conflicts, to express them. On stage you could do anything. I could get on the stage with Margret and say, 'You asshole, you slut, I hate you.' And you could say it sincerely. And there would be no reprisal. And suddenly with Margret as the theater director you couldn't. There was a clampdown. So even the theater totally changed into some ideological dictatorship."

I will always wonder if John and Margret saw our play as prescient or bespeaking our state of mind at the time, and whether it played a role in the sudden dismantling of our team.

With the looming deadline of Closure, and so much still to be done, with hostile media and CO_2 problems, pressure mounted almost daily. John did not seem to be taking it well. He was becoming erratic and increasingly authoritarian. After-dinner discussions were getting cantankerous and dreadful, and often John would give someone a verbal lashing.

One day Linda had stated a concern over a soil fungus called *pythium* that can attack plants. It was a legitimate issue, but apparently our Biosphere was going to be perfect. That evening John declared that Linda was "negative" and made her jump up and down in front of everyone, yelling "pythium."

Safari recalled that, "Some of the Thursday nights before Closure were really scary. He [John] was making the atmosphere so dense that you could cut it. And people were so fearful that they would be pointed out next. Once he threw a plate at me because 'Safari is overgrowing the Agriculture with loofahs.' He got so into it and so angry that he threw a plate at me, which missed me fortunately, and he told me to go right now and dig up all the loofahs. And I did. And then I came back and was all happy, as we did. I said, 'Thank you, that was great!' " John was our wise patriarch whom we all needed to impress.

I began to dread seeing John.

We all dreaded Margret's call on the two-way radio, her deadpan voice summoning, "Harlequin, Harlequin, do you read me, over," intoned as if she was announcing the time, or the weather. Butterflies would always flap violently in my stomach, my heart pounding as I threw down whatever I was doing and trudged to her office in Mission Control. It could only mean one thing—I had screwed up royally.

Some called this era in the project's history "the reign of terror," with John and Margret playing king and queen with a siege mentality. So much for our open, high-minded society.

I had developed a report to show how our crop harvests were holding up against the yields required for us to eat comfortably during our two-year confinement. It was a tricky report because the glazing had only recently been completed. But I used a combination of historic yields and judgment based on best possible yields from the research greenhouses. I was projecting that we would be able to grow about 80 percent of our food in Biosphere 2.

John had sat with me on a couple of occasions for previous reports and shown me how to bring the total yields up to 100 percent by including imaginatively optimistic yields from shaded, immature fruit trees, and other crops. He had done the same thing to Safari when she had been in charge of the Agriculture design; that was why she had been happy to dump it on me.

It was time for me to produce another report to present at our weekly biospherian meeting, and I was still showing about 80 percent production. I knew it spelled disaster, but I inherited my father's insistence on following certain principles of ethics, so I simply could not bring myself to write a phony report.

I announced to Bernd, our captain, that I would not bring the report to the meeting. He agreed. I sat with six other biospherians around the oval oak table in the Mission Control conference room, with John in the middle, his back to the door. (Gaie was away at the time.) Finally we came to the report. My heart was pounding as I announced, "I don't have it."

John leaped to his feet and ran out of the room yelling, "I'm resigning!"—not for the first time. The door slammed and we all sat in shocked silence. For some reason we kept sitting. We sat for an eternity until Margret entered and declared that Bernd, Gaie, and I were fired.

I was thunderstruck. It felt like my heart hit my feet. I had been in the inner circle, one of John's closest allies. John had forced me into the position of defying him. I had witnessed Safari yanked from the crew, and now I was anticipating talking to Taber through the glass for two years, and wondering what I was going to do with myself on the outside. That night I cried with Taber.

Three days later Margret called me to her office; I thought we would discuss my new role. Instead, she informed me that I would, after all, be going in the Biosphere. I have never been more relieved.

Nothing about Gaie was ever mentioned again. It was puzzling that she had been fired at all, as she had had no hand in the incident—she wasn't even there. Some speculated that it was simply to strike more fear into our hearts, a random act that could happen to any of us at any time. Gaie, too, remained on the crew. Bernd eventually told the press that he had a health problem.

Margret announced that Sierra was now captain of our team. I knew little about Sierra. Even after the years we would spend cooped up together, it is strange that I know almost nothing about her and some of the other biospherians' pasts. I know what is on their curriculum vitae but not how they grew up, or even how they found their way to IE and the Synergias.

I know that Sierra is English and was thirty-six when we entered Biosphere 2. I know that for a year she worked in Kenya with orphaned children, and in India for a couple of years, developing agriculture programs (but I didn't even know that until I looked at her resumé). I know from her that she spent some months working in a signal box along an English railway in her late teens or early twenties.

But that's about it. We just didn't talk about our lives before, our lives outside the nonstop excitement of the Biosphere 2 project.

Sierra had almost-black straight hair, cut in a twenties-style bob with pointy sideburns and all, which I would manage to keep vaguely intact throughout the two years (I was to be the not-very-trained hairdresser). She had a bony, but not unattractive face, atop strong shoulders and a large, round body. She vigorously chewed the end of a ballpoint pen during meetings, and woe unto anyone who accidentally picked up Sierra's pen—spittle would drip from it.

Mark Nelson, chairman of IE, was to replace Bernd on the team. He had become a candidate only a month earlier and had hardly been on the project, but was to be in charge of communications and waste recycling. A small, wiry man in his early forties, Mark's full head of short-cropped hair was graying and always looked scrunched up. He was a Dartmouth graduate and had worked like a fiend on several of the IE projects, particularly Savannah Systems, the seed-production project in Australia.

He told horror stories of fires blazing across the savannah. One particularly ferocious fire headed straight for the project, and he and five or so other people struggled all night to save the property. They lit fires to burn swaths of land between them and the wall of flames galloping towards them. They hoped the wind would not change direction and turn the fires they had set back their way. The firebreaks worked and they won.

And the word *wattle* held particular significance for him. When IE first took possession of Savannah Systems, there was little open pasture for planting. It was covered in an invasive species of small tree called wattle. For years the small crew spent hour after back-breaking hour manually chopping down the wattle to reclaim the land for grasses.

Eventually they used bulldozers for the rough clearing, but still needed to manually chop out the roots, which would otherwise regrow.

I chopped wattle for two weeks on my way to Quanbun—it is damn hard, sweaty work. It was this extraordinary work ethic that Mark brought to our team, though he was taciturn and almost always appeared harried.

I felt awful for Bernd. He had done an incredible job of managing the Biosphere's construction. Mostly, I felt bad for us, the rest of the team. He was jovial and a good leader. It was a tragedy that we lost him. Bernd never tried to get back onto the team, which, according to Synergia lore, he might have been able to do had he declared with enough intensity to John and Margret his intentions and determination to be the best biospherian the world had ever seen.

Bernd later explained to me, "One reason I got kicked off the biospherian crew was I defended you. That is something that was not highly regarded in this culture. I think our play had something to do with it, too. We predicted our future."

A couple of staff scientists defected, and one was seen on national television claiming improprieties. We were supposedly hiding a CO_2 problem and altering data. Although John had displayed a bent toward being optimistic with projections, the claims of fraud were baseless.

Despite the bad publicity, hordes of reporters were beating down our doors to peek inside before the Biosphere was closed up and they could no longer explore.

As spring yielded to the intense heat of summer, we hammered away at checklists, operating procedures, and emergency drills.

The wave machine in the Ocean would not pump and the Ocean turned pea green with a nasty algae bloom. The rice-paddy irrigation system ruptured and filled the freshwater tanks with mud. The analytical lab equipment was being finicky about the hydrogen gas it ran on, which was generated from the water inside the Biosphere. The global monitoring system refused to give data. The radio alarm system went down. A hot-water pipe leaked red water into the Biosphere. The Energy Center kept breaking down and all power went off.

But one by one, the problems were solved. Countdown had begun.

So why were we pushing so hard to start a two-year mission when clearly it would have been better to have a six-month maiden voyage? Money became an issue. The original cost estimate given to Ed Bass had been for about $30 million, though apparently Ed had been wise enough to figure on three times that. The project had grown in scope, we had missed deadlines, and now the cost was passing $100 million. It would be much more before the Biosphere was complete.

From what I heard, Ed tried to get banks to loan the project the remainder, but failed. Eventually he struck a deal with SBV. He would spend $150 million to Closure and no more. Once the two-year experiment started, an operating budget would kick in, and money could start flowing again. Finish the Biosphere. That was Ed's message.

At the time, I was almost entirely oblivious to money matters. My world was 3.15 acres inside the dome. I was more interested in the first biospherian birthday cake, made and eaten on May 29 in honor of Mark Nelson. On June 2, Linda brewed the first cup of Biosphere 2 java. Closure was becoming more real all the time.

On July 8, the first animals were born inside Biosphere 2, six healthy piglets. We were using Ossabaw feral swine, small pigs that had been brought over to the U.S. by the Spanish and turned feral along the East Coast. Ossabaw Island, off the coast of Georgia, was the only place known to still have the pure breed, thus the name. They were extremely hardy pigs, able to turn almost anything that remotely resembled food into meat and fat.

This pig was our second choice, which we had turned to once our first choice, the Vietnamese potbellied pig, had to be removed from our domestic animal list. Smaller than the Ossabaw feral swine, the potbellied pig is used all over Asia as a farm animal, with an amazing ability to turn stuff that does not even look like food into fat. Our diet was going to be low in fat, so this was a big plus.

Unfortunately, the potbellied pig was in its heyday of petdom in the U.S. Darling little piggies were being clothed in pink satin dresses and taught how to use and flush toilets. When the animal rights movement got wind that we were going to eat their pets, they sent threatening letters to Margret. So, Phoebos and Demos were replaced with Quincy and Zazu as the breeder pigs in Biosphere 2. Zazu's six offspring were the ones squealing bloody murder in the animal bay.

Each Reality Experiment came closer to life inside Biosphere 2. Coupling Reality began on August 15, when all the biospherians slept, ate, and worked inside Biosphere 2 with only a skeleton construction crew working inside. For the first time, the water cycle within Biosphere 2 was completely closed. All the water we were drinking was being transpired by the plants as pure water and then condensed by heat exchangers. After we drank the water, it went into the waste-recycling system where it was treated, then irrigated the plants that transpired the pure water into the atmosphere.

We were continuously drinking the same water. Upon hearing that, many people exclaimed, "Ooh, yuck!"

I, on the other hand, thought it was thrilling. It was a biospheric cycle. It meant we had an essentially hermetically sealed system, a step closer to a self-sustaining biosphere, independent of Biosphere 1 for anything but energy and information. This was also the first full test of the waste-recycling system, which was taken from a design by Billy Wolverton at NASA. Using what is now considered fairly standard biological waste treatment, the waste is first sent to anaerobic settling tanks before being pumped to a marsh system where plants and microbes break it down and absorb the nutrients until the water is clean enough to use for irrigation again.

Our system seemed to be holding up, even with all the camera crews inside, although they were not supposed to be using our facilities. Piece by piece our new world was beginning to work.

At our daily morning meetings, the CO_2 level was given as part of the weather report. On August 17, the highest level was 1,300 ppm, recorded in the morning before the sun on the plants began to bring the level down again. Taber was also detecting large amounts of trace gases being emitted by the plants, which was not a problem, but a sign that his analytical systems were working well. Taber and Roy took blood samples from each biospherian for their blood-toxin research, which would look at what chemicals were stored in our fat. The compounds would be released from the fat in the spare tires around our midriffs as we lost weight.

August 18 was the first day that biospherians inside Biosphere 2 outnumbered the construction crew. We all moved the few belongings we were going to take with us inside. But how do you pack for two years? I had been performing shoe trials to see how long my work shoes lasted—six months was the longest, so I took in five pairs. I packed an extra suitcase of clothes that were a size too small, that I would open at the end of the first year, knowing that we were bound to lose weight and that I would be dying for something new to wear.

Taber and I headed into town, figuring it was going to be one of our last chances to get anything we needed. We went to an Albertson's supermarket on the other side of Tucson from the Bio-sphere. We grabbed a shopping cart and started stuffing it with every bottle of booze we could find. If we were going to be stuck inside Biosphere 2 for two years, at least we could celebrate a little along the way. After all, it was the tradition of the British Navy to give their sailors a tipple of grog a day, and the caloric input would be insignificant.

Once our basket was full to overflowing with bottles of all sizes, shapes, and colors we headed for the cash register. I almost jumped out of my skin when I heard Margret's unmistakable laugh and saw the back of her curly head bobbing down the neighboring aisle.

How did she know we were here? She could not have, but she had an uncanny sixth sense for finding any of us doing something we should not be doing. We left the basket and ducked out of the shop, hoping she had not seen us. We never got the nerve to try again, so we had to mostly make do with what we could make.

On August 29, we were back on our biospherian diet when we began a weeklong full-up simulation of Closure, beginning with walking through the airlock and closing it behind us as we would on September 26. It was a sunny day, and the CO_2 was driven as low as 480 ppm that day.

August 31, and the CO_2 low was 554 ppm. It was rising, and the soil-bed reactor had already been turned off. On September 1, clouds darkened the Biosphere and the lowest CO_2 reading that day was 800 ppm. Apparently the Biosphere's CO_2 level was extremely sensitive to light input. September 3, the CO_2 low level was 986 ppm. The CO_2 was still rising. We had a problem.

Taber had prepared a briefing for John that proposed installing a carbon dioxide scrubber similar in nature to those used on submarines. John was initially adamant against any form of nonbiological life support, but eventually had to agree that it would be a good insurance policy, only to be used in an emergency.

On September 16, only ten days before we were scheduled to enter the Biosphere for two years, Taber began building a regenerative, physical/chemical method of scrubbing CO_2 out of the atmosphere of Biosphere 2. Taber had not been able to buy one, so he designed a system that essentially turned carbon dioxide into limestone using sodium and calcium hydroxide. To liberate the CO_2, the limestone simply needed heating.

Along with active maintenance of soils and plants so that the plants were absorbing as much CO_2 as possible and the soil releasing as little as possible, Taber calculated that the scrubber would remove enough in the winter to prevent the level reaching a point where it would lower the Ocean pH to dangerous levels, or affect our health. It could then be recycled into the atmosphere in the summer when the plants would need it. Taber and a crew of about fifty people worked to install the scrubber, which they finally finished and tested on September 25, the day before we were scheduled to enter.

During our Closure simulation, Michael Eisner, the CEO of Disney, toured the Biosphere along with Ed Bass and Ed's older brother, Sid. After the tour we all stood in the Biosphere 2 dining room, biospherians munching on chapatis, beets, and mashed sweet potato, the dignitaries slurping delicacies from the on-site restaurant. We were all craning to hear what Michael Eisner thought of the place. "Well, it's fantastic. You've only made one mistake," he said, and we wondered what technical gaffe we had made.

"You didn't build it in Orlando!" Everyone laughed politely, but with tourism now a major contributor to the bottom line, he was right.

One evening a few weeks before Closure, I sat cross-legged on the "mountain," the hill along the north and west sides of the Biosphere with John, Tango (a Synergist who ran a publishing company and produced the Biosphere 2 publications), Gaie, Laser, and Taber. The full moon made each cholla needle fluoresce, and below, the almost completed Biosphere seemed to have come alive, a luminous sphinx crouching in the valley ready to pounce into an unknown future. She was absolutely beautiful, and I was in love with her.

With expansive youthful exuberance I swore to take care of her during our two-year journey together. Closure was near and it signaled a new beginning. I put all my misgivings and conflicts behind me. Somehow we had built Biosphere 2. The six of us howled at the moon, and all but drew knives across our palms in a YaYa blood pact, swearing that nothing was going to come between us, pledging allegiance to each other, to Biosphere 2, and to the successful completion of the two-year enclosure.

Soon after, the same six went out for a celebratory dinner, to our favorite hangout, Le Bistro. A steaming bowl of mussels in white wine would provide our last supper before Closure. We only had a couple of weeks to go until we were to be ripped away from our high-fat, high-sugar, high-caffeine diet, and imbibe nothing but the wholesome, homegrown, organic, really-good-for-you-but-not-always-tasty food of Biosphere 2. We reveled in our camaraderie, the rich French food, and fine wine.

We played our routine prank on George, the Belgian owner, inviting him over for a drink after the waiter had lowered the lights so that he would not see the cellophane covering his glass. We did it every time we dined, he never suspected it, and we always collapsed laughing at the look on his face when his nose bounced off the invisible barrier on his glass.

La pièce de résistance came when George removed a threatening saber from a leather case, and with great ceremony sliced the top off a champagne bottle and poured it over a pyramid of champagne glasses. In high spirits we drank to a successful mission: All for one, and one for all.

Meanwhile, the Biosphere cycles of life turned relentlessly. Every morning the goats needed milking and all the domestic animals needed feeding. Taber and I butchered two of Zazu's piglets. Crops needed watering. Plants and animals in the Wilderness biomes needed tending. Troubleshooting continued.

We were preparing to enter our materially sealed world. The eight of us were to stay inside Biosphere 2 and not come out for exactly two years unless one of us had a medical emergency, in which case the afflicted person would exit via the airlock. During the two years, nothing was to enter or leave our hermetically sealed environment—no food, no water, no air (except the less than 10 percent that leaked out), no junk mail, no new CDs, no new clothes, no parts to replace anything that broke—nothing.

However, we would be able to communicate with the outside world. We had telephones, e-mail, videoconferencing with Mission Control, and primitive videophones to talk with people on the outside who also had them. We had satellite TV to watch and radio to listen to.

And we could see colleagues, friends, and family at the visitors' window next to the airlock where we could talk to people on the outside via phone. We would stand inside a four-by-six-foot curtained room in front of a big picture window that started at waist height. Whether it was rainy, snowy, or blazing sun, the person outside would stand unprotected from the elements, talking with us by telephone. It would be a far cry from touching the person, but it was something. We would at least be able to look a friend in the eye and shake hands biospherian style—one hand covering another, separated only by the cold glass. In the two years to come, lips would also leave their imprint.

Power could also penetrate our world. Our Energy Center and the grid power were to be in some ways what the sun is to the Earth—its energy source.

In short, we were frantically getting ready to close the door on a materially sealed, but informationally and energetically open world. In this regard, Biosphere 2 was essentially identical to Biosphere 1.

As we prepared for the great celebration and Closure ceremony, we got media training—the art of responding to any question with a prepared sound bite, no matter how oblique it may be to the original question. So, for instance, if I was asked "Is there going to be sex in the Biosphere?" I would phrase the answer to talk about the virtues of growing many different crops at the same time in order to have a complete diet. It was like a game—could I get all my sound bites out and say nothing more, no matter what I was asked? I hate it when politicians use this method of interviewing, but I am no politician. I just did not want to put my foot in my mouth, be misquoted, or have my words taken out of context.

My parents arrived. They were adoring and adorable. They brought me a present for each birthday and Christmas. They brought Christmas decorations for us to hang around the Habitat during the holiday season, including Christmas crackers—a British tradition. Each cracker is a cylinder of cardboard about five inches long, wrapped in foil that bunches at each end. Two people grasp each bunch and pull. The cracker breaks with the sharp crack of a tiny gunpowder explosion, and silly hats, jokes, and trinkets fall out.

CNN interviewed my parents, my beaming mother repeating her rehearsed line, "I'm so proud of Jane!" She was proud. And afraid I might fall ill or be hurt. And happy that she would know exactly where I was for once.

Two nights before the big day, a Crow Indian medicine man, Dan Old Elk, held a sweat lodge for all of us to cleanse our spirits. The day before Closure, Sierra and I held a satellite uplink, fielding questions from the world's media, while most of the rest of the crew participated in an environmental symposium chaired by Claus Nobel, a member of the Nobel Prize family and founder of United Earth. Hundreds, thousands of people flocked to the sight in the desert. We signed autographs on pamphlets, napkins, hats, and sweaters, and judged paintings done by kids from all over the country.

During a press conference that afternoon, the catering truck holding the food for the evening's party caught on fire. Black smoke billowed out the doors and windows. A van carrying emergency food sped from Tucson with a police escort just in time for the huge party that night.

That evening, two thousand people mingled on the lawn in front of Biosphere 2, watching fire jugglers and African Spirit dancers on stilts who somehow squeezed themselves through the airlock door into the Biosphere. We had an inspirational phone call with author Arthur C. Clarke. Woody Harrelson and other actors from the TV show *Cheers* entertained us. I am sure I was supposed to gush with excitement when I met them, but I did not have a clue who they were, not having watched TV in years.

The grand finale was a laser-light and fireworks display over the Biosphere. It was one hell of a party. People had come to send us off who I had not seen for years. We hugged and kissed, and cried, until finally I said goodbye to my family, knowing that the following morning would leave no time for sentiment.

Meanwhile, in the basement of Biosphere 2, Peter Warshall was averting a crisis. The galagos were not yet able to completely sustain themselves on the fruit inside the Biosphere because the trees were not mature enough. So Peter had supplied Purina Monkey Chow to supplement their food. During the festivities, he had wandered down to the Biosphere 2 basement to do a last-minute check on the galago nosh and discovered that it was infested with worms. So, while everyone else was celebrating, he emptied every bag of food and baked it in the Agriculture drying oven to kill the worms and save the monkey fuel.

Caravans of media had arrived. By the time we got up on the morning of September 26, vans lined the hill behind the Biosphere. The day before, Kathy had panicked when she finally heard that a CO_2 scrubber had been installed. She had already printed press releases in large quantities. It was too late to include it in the story.

She knew that people like Bill Broad of *The New York Times*, who had just written a lengthy article comparing the Biosphere 2 systems to NASA's approach, would feel betrayed by not being told about the inclusion of a physical/chemical system (despite our spinning it as our biospherian volcano). There was nothing for Kathy to do but hope like hell no one caught wind of it, though she knew that was unlikely.

The eight crewmembers gathered on the lawn in front of Biosphere 2 for a blessing ceremony with Dan Old Elk, a Mexican sorceress, and a Tibetan lama. It was a beautifully simple ceremony after the grandeur of the evening before—a stringed instrument, dancing and chanting. Finally Dan Old Elk said, "Turn and face the sun, because it is most important to you. I will perform the Sundance ritual for you." He would hang himself from a tree by hooks through the skin on his chest, an ancient Plains Indian ritual.

A few people from the computer systems were finishing last-minute details while the eight of us stood on the platform outside the airlock staring down at an ocean of cameras and friends. Ed, who rarely stood in the spotlight, stepped up to the podium and, with hundreds of cameras and millions of eyes on him, began speaking. "When Galileo built his first telescope, more than four hundred years ago, he did not know what he would see when he peered through it. But he eagerly anticipated that he now had an apparatus to study the heavens as never before possible; that through his new instrument some theories of old would be disproved while others would be confirmed; and that never-before-imagined mysteries of the universe would be revealed. And so the modern age of science was born.

"More than four hundred years ago, Leonardo da Vinci imagined machines that would fly and carry men aloft like birds. Hundreds of years later, the Wright Brothers launched their first successful airplane at Kitty Hawk. Others had tried before and failed. . . . The Wright brothers, with typical America ingenuity, succeeded with their first flight that today seems ever so tentative and fledgling, and the space age was born into infancy.

"Like Galileo, we eagerly anticipate our discoveries, though we know not for sure what they will be. Like the Wright brothers, we are hopeful that our efforts will soar, though we are prepared to meet our failures as well. We will have success, I am confident of that; and we will have failures, I am equally confident of that. We will meet the challenge of both with ever harder work and new doses of ingenuity.

"We strive, ultimately, to build a machine that flies as well as our own beloved, living, spaceship Earth. With the perspective of time, our efforts will certainly seem as tentative and fledgling as Kitty Hawk. But with the experience of time, I hope that Biosphere 2 will make an even greater contribution by changing the way that we view our planet and the role of humankind upon it—by carrying us into a new age of awareness.

"Biosphere 2, as we see it before us today, is meant to be beautiful, accessible, and very real. It has a tremendous capacity to help us become more aware of what makes our planet tick, to make us appreciate that all of humankind lives in a biosphere, and that the role that humankind plays is pivotal. We can be a threat to our biosphere, playing a destructive role; or we can be stewards, contributing to the reciprocal maintenance of natural ecological processes.

"Today, eight biospherians enter Biosphere 2 with the exact same potentials—and they will strive to be good stewards. I envy them the richness of the life that they will lead, with a whole world of beauty and nature in a moment's reach; I envy them the excitement of discovery that will be a part of their day-to-day experience.

"Biospherians, good friends, bon voyage, and fly your spaceship well, that all of humankind might fly its spaceship Earth better in the future."

We gave brief speeches, moving the crowd to cheers, and some to tears. The call was given over the two-way radios that we had finished and were walking toward the Habitat airlock. Final stragglers fled out the Savannah airlock on the opposite side of the Biosphere. Amid camera shutters clicking, cameramen ducking and diving to get the right angle, we lined up at the airlock, waving as we entered, breathing the last few molecules of Biosphere 1 atmosphere. At eight o'clock in the morning on September 26, 1991, the airlock door slammed shut behind us, and we opened the inner door into Biosphere 2, to silence. We were alone for the first time in the world's first hermetically sealed, man-made biosphere. The CO_2 was low, below ambient at 318 ppm. That was auspicious.

The Dalai Lama has advised, "Take into account that great love and great achievements involve great risk." We had taken huge risks in building Biosphere 2. We had broken rules, done things never before attempted, tested friendships to their breaking point, and made enemies. But in spite of it all, we had done it. We had come this far. We had built Biosphere 2. We had done the impossible.

"None of us is as smart as all of us," write Warren Bennis and Patricia Ward Biederman in their book *Organizing Genius, the Secrets of Creative Collaboration,* referring to the power of a group to accomplish things. They believe that the end of the great individual, heroically battling adversity alone, has arrived—the world is simply too complex for a single person to tackle many of today's challenges. "That does not mean that we no longer need leaders. Instead, we have to recognize a new paradigm: not great leaders alone, but great leaders who exist in a fertile relationship with a Great Group. In these creative alliances, the leader and the team are able to achieve something together that neither could achieve alone. The leader finds greatness in the group. And he or she helps the members find it in themselves."

Our group, however unorthodox, and perhaps because of its unorthodoxy, was such a Great Group. Kathy Dyhr describes the group that built the Biosphere thusly. "There was a group of talented people wanting challenges, willing to think unconventionally and outside the box, willing to devote themselves and to work for hardly any pay, just enough to get by, and to work like madmen. We did—twelve, sixteen hours a day, seven days a week. Everybody put their heart into it. If something was falling down, someone else would come in and pick it up."

However, the pressure of realizing the dream of building the world's first man-made biosphere had ripped the sails that propelled this Great Group. We had patched the rifts, but would they hold long enough for us to sail together through the two-year maiden voyage of Biosphere 2?

As the doors closed behind us, I was exhilarated by the crowds and cheering. I was relieved I hadn't flubbed my speech. I was thankful I hadn't done anything dumb like trip over the foot-high threshold into the airlock. I was exhausted from the schedule and pressure. But I was happy, oh so happy, to be finally setting sail on this journey that we, along with hundreds of other people, had worked tirelessly to make happen. I was chuffed. No, I was bloody proud.

But I also had a mild sense of foreboding, knowing that building Biosphere 2 was only part of the picture. Now we had the daunting task of operating it, of testing whether our creation would work. All our attention had been on this moment—finishing Biosphere 2. Were we ready for what came next?

Chapter 10

WHAT HONEYMOON?

All eight of us crammed in front of a window next to the airlock, waving to the world outside waving back at us. Cameramen and friends were elbowing each other to record the opening moments of eight people locked inside a hermetically sealed biosphere for the first time in history. Then we walked away from the dazzling attention and up the purple-carpeted spiral staircase to the black-and-white kitchen for a cup of homegrown mint tea.

As we passed the analytical laboratory, Linda glanced through the porthole in the lab door and gasped, "What is Nina still doing in here?" Nina, one of Taber's technicians, had worked all night in the lab to finish up last-minute details. Taber blanched. He looked through the porthole, but relaxed, laughing as he saw the lab was empty. Spirits were high. A sense of euphoria had settled over the crew as we embarked on our journey.

I spent most of our first day inside weeding in the Agriculture, feeding the domestic animals, and milking the goats. That evening, I strolled around the Biosphere to take stock of this new world. Although I knew almost every nook and cranny intimately, and could have walked it with my eyes closed, I toured our home with fresh eyes. This was our Biosphere, population eight. For the first day since we began construction, the rest of humanity was outside in the other biosphere, Biosphere 1.

I left the animal bay, where the chickens, pigs, and goats crowed, snorted, and bleated right under our bedrooms. I marched across the ground floor of the Habitat, passed the entrance to the Intensive Agriculture biome and the machine shop, and swung open the heavy door to the orchard, where guavas, coffee, star fruit, Anna apples, and citrus grew.

Stuffed in between the Wilderness and Agriculture biomes, the small orchard was shaded in the morning by the Savannah, and in the afternoon by the Agriculture, so the light was poor. But I saw a few flowers bursting with pollen on some trees. Stairs led up to a platform outside the dining room, and through a glass door into the Savannah, which was a full two stories higher than the Orchard and Agriculture.

But I followed the path to another staircase and descended into the gray cement and steel basement, the Technosphere. I opened the door to the sound of humming air-handler units that controlled temperature and humidity. Encased in room-sized metal boxes, they stood along one wall of the Savannah basement.

I turned right, away from the Rainforest basement and downhill towards the Desert. The red lights on the sipper system flashed, and I could hear the pshshshst-plunk as it took a sample. Tubing ran from the Ocean and the other water bodies in the Biosphere to a series of sensors and analytical gadgets that tested the water automatically for parameters like pH, salinity, nitrates, and other nutrients, vital health signs for each area.

I passed air handler after air handler, Ocean pumps, and a flash evaporator for reducing the Ocean salinity if needed. I glanced up and marveled at the neat rows of hot, cold, and chilled water pipes and electrical and sensor conduits lining the concrete ceiling. The basement led on under the Desert, past irrigation-water holding tanks and pumps and more air handlers, until it narrowed to a small tunnel, just tall enough to walk through, that sloped down to the South Lung.

The Biosphere had two of Freddy's lungs, one attached to the south end of the Wilderness, the second one attached to the Agriculture basement on the west side of the Biosphere. It was both awe-inspiring and spooky to walk into the center of one of the 158-foot-diameter circular structures, and stand under sixteen tons of metal pan hanging ominously overhead, attached to the steel walls by a flimsy synthetic rubber skirt and held up only by the pressure of the air inside the Biosphere.

The acoustics in the West Lung were extraordinary, sound bouncing off the metal floor and walls in seemingly ever-cycling echoes. Tibetan lamas had chanted in the lung. Paul Horn played flute in the lung.

But the South Lung housed two hundred thousand gallons of water, overflow from the irrigation system that could be used in case of fire, which dampened its acoustic effects. The red hydrant in the center of the reservoir stood out starkly, the only color in this gray underworld. On this, the first day of our two-year mission, the lung was about 50 percent inflated, where it should be on a day that was not very hot.

I retraced my steps to a small door wedged between two air handlers in the Savannah basement. Wind hit me as I entered the plenum, a narrow walkway that funneled air from the air handlers to outlets over the various biomes. Making sure the door was closed behind me so air would not circumvent its usual passage, I battled my way through the machine-made gale, up a short flight of fiberglass steps and through another door.

The noise was deafening as I walked into the room where three air pumps ran continuously to make the artificial ocean wave. The pumps controlled actuators that sucked around ten thousand gallons of seawater into a hollow wall that ran the length of the south end of the Ocean, only to precipitously drop the water, causing a wave to flood across the reef. The rhythmic surge allowed the static coral polyps to grab little pieces of food with their tentacles as the edibles drifted by. Without the wave the corals and their relatives would not receive their meals and would die.

I hurried into the scrubber room, closing the door behind me and was greeted by the sound of rushing water and the smell of the beach on a hot day, seaweed stinking under hot lights.

Walter Adey had devised the ingenious algae scrubbers, which he had tested at the Smithsonian and elsewhere. The water around a thriving coral reef is very pure, with only low levels of nutrients. Many of the problems assaulting coral reefs around the globe today stem from too many nutrients, as well as pollution and sediments washing over them from nearby rivers. Industry, poor agriculture practices, sewage, topsoil loss from deforestation, and coastal development all contribute.

Adey's theory was that, under normal conditions, thousands of square miles of open ocean purify the water that feeds the reef. The phytoplankton—photosynthetic microorganisms—fill waters where light penetrates, absorbing CO_2 nutrients, and other compounds. The open ocean also plays an enormous role in the Earth's carbon cycle by absorbing and storing huge amounts of atmospheric CO_2. In Biosphere 2, the carbon cycle could alter the pH of the mini-ocean if the atmospheric CO_2 level rose significantly, possibly killing many sea creatures.

So Adey condensed thousands of square miles into less than two thousand square feet. High intensity lights shone onto plastic mesh where many species of green, brown, and red algae grew, stripping nutrients and CO_2 from the Ocean water that washed over them. Once a week we took turns scraping the algae from the mats with a piece of plastic resembling a wide paint stripper, bits of errant algae spattering our legs and arms. We wadded the algae into what resembled sloppy green cow patties that we dried in an oven. This stinky, laborious task took each of us a couple of hours every week, allowing room for more algae to grow to continue the cleaning cycle.

From the scrubber room I emerged at the bottom of a fake rock face with steps cut into it leading up to the Thornscrub, the *ecotone,* or merging of the Desert with the Savannah. Before heading up the stairs, I popped in on my friends, the mangroves that had caused so much trouble at the Arizona border.

It felt just like standing in the Florida Everglades, but without mosquitoes or the threat of an alligator attack. I briefly hunted around for a garter snake, a venomless black snake with fine blue and red stripes running the length of its body. The snakes would feed on the mosquito fish in the freshwater Marsh and the stream that ran the length of the Savannah, keeping their population in check.

At the top of the steps I looked out over the Desert twenty feet below, the southernmost point of Biosphere 2 aside from the South Lung. The Desert seemed isolated, a long way from everyone, a good place to get away from it all. I turned and picked my way along the path through the Thornscrub, careful not to be stabbed by the spiny plants, then strode across the open plain of the Savannah. The grasses lay dormant now, dead leaves heaped over each other.

I stood for a moment at the edge, looking down the three-stories-high cliff face into the mini-ocean, one foot resting on the concrete rock ledge next to a frankincense plant given to us by the Sultan of Oman. In its natural habitat, frankincense flourishes in limestone cliffs in deserts with summer monsoons like our miniature Savannah, so the plants flourished in the cement pockets of the cliff.

"The biospherians were not to eat out of the Wilderness, but it did not mean that they could not use small packets of value, like medicines and spice," Tony Burgess says. "We used the term *psycho-diversity*—all the plants went through an additional filter of medicinal and condiment uses, thinking they would likely be better cared for than useless plants."

The ancients believed frankincense's fragrant smoke carried prayers to the gods. They used its essential oil in perfume. Early doctors thought it cured depression, claustrophobia, eczema, and abscesses. Egyptians embalmed mummies with it, and because of its skin-preserving quality, the living used it as a cosmetic. At the trade's peak two thousand years ago, the Dhofar region of Oman reportedly exported up to three thousand tons of frankincense each year, priced as high as gold. For a time, Biosphere 2 grew the only frankincense in the U.S., but we gave cuttings to botanical gardens around the country.

I jogged over to narrow steps that wound down the cliff face, reaching the beach, where I sat under the coconut palms. They would not bear fruit for another ten years, too long to do us any good, but they would provide delicacies for crews to come. This was my favorite spot in the Biosphere. It felt almost normal to sit on our beach. I could see fish swimming in the gentle waves lapping up onto the shore. I wondered if those were ones I had caught. Overhead arched the geometric patterns of the white spaceframe, through which I could see the red-splashed clouds and the crimson mountains of the Catalinas in the setting sun. It did not take much imagination to think I was nestled on the red planet Mars.

The sound of tapping on the glass shattered my thoughts. I looked up to see a group of onlookers waving and snapping photos at the window. I waved back, and left via the stairs to the Rainforest, to the seclusion of its narrow path in the shade of the tall upper-story trees, to the foot of the thirty-foot-tall mountain in the center of the forest. The glass-and-steel roof amplified the sound of the water thundering down from the mountaintop into Tiger Pond.

As I wandered along the path, I tried to imagine what the immature forest would look like in ten years, twenty years. Sir Ghillean had hypothesized that the Rainforest structure would change from that of a recently cleared area to a complex primary forest. I imagined hacking through dense undergrowth with a machete to reach the mountain.

Instead, I charged easily along the path and through the entrance into the hollow mountain, down the spiral stairs to the basement, past more air handlers, through more doors, and down more steps, finally turning right at the Savannah airlock into the Agriculture basement. I could not help thinking what a maze this place was.

Large fiberglass tanks lined the walls of the basement next to the windows on the south end, taking advantage of the light to grow rice. It was almost ready to harvest. Next to those were the waste recycling tanks, growing canna lilies and other marsh plants that we could periodically harvest above water level and feed to the animals. This was my kingdom. I knew it intimately, so I did not linger long, knowing I would be back soon. I ran up the stairs, out the door behind the banana trees growing along the north wall of the Agriculture, and back to where I started, the ground floor of the Habitat. Before heading up the Habitat stairs I took off my shoes, as had already become our custom. Plodding along, it began to dawn on me how large 3.15 acres are. A whole biosphere needed our tender loving care, and the eight of us really were the only people inside it.

Reflecting now, I see how beautiful our life support system was. It was visually stunning, with glorious architecture and majestic plants absorbing the CO_2 and producing precious oxygen. And it was technically beautiful, with the mechanical and life systems working in concert to maintain a functioning biosphere. Conceptually it seemed that technology and humans had found their rightful place, working hand in hand to maintain and enhance life, not damage and destroy it. In 1991, the thought that humanity could either destroy or enhance life on a global scale seemed crazy to most people. Now we know differently. The threat of largely human-caused global climate change has even been acknowledged by the Pentagon.

At the time, though, I was thinking about food. My stomach was growling, and I hoped the dinner bell would soon ring. And there it was. Over the two-way radio came the clang of the Tibetan bells.

Already trained like Pavlov's dogs, my mouth immediately started to water. Time to fill up on some calories.

After a good night's sleep, we slipped immediately into the daily routine. Up at six thirty, we all spent an hour before breakfast in our areas of responsibility. Sierra and I fed the animals, milked the goats, collected any eggs that the chickens might have kindly laid, and cleaned the pens. Then I would check the crops and irrigation in the Agriculture. Linda would start her daily chores in the Wilderness biomes, including making sure she had enough water to irrigate with that day. Gaie would check that the algae scrubbers were functioning properly, that the wave machine was working and that the Ocean pH was between the required 8 and 8.4. Laser did the rounds of the mechanical systems. Mark would glance at the waste recycling systems and then harvest animal fodder. Sierra surveyed the Agriculture and food systems and prepared for the morning meeting. Taber ensured the automatic monitoring systems were functioning and then helped Linda in the Wilderness, and Roy prepared for the day's medical exams or helped in the Wilderness.

Over a half-hour breakfast, Mark read the day's weather report, including high and low carbon dioxide levels for the last twenty-four hours, and we would discuss the day's schedule and priorities. All eight of us worked in the Agriculture for the first two hours, where I would boss everyone around, getting crops planted, weeded, or harvested, or the soil turned. At ten thirty I bellowed, "Break time" at the top of my ample lungs; everyone charged into the plaza outside the Agriculture at the base of the main staircase. Sierra would have laid out eight neat handfuls of peanuts and the first person got the pick of the piles. Everyone would lean over the heaps studying each one in detail to grab the handful with the most peanuts. Then we ate them shell and all, just to put something into our whining stomachs; a cup of mint or lemongrass tea washed them down.

Then Sierra, Roy, Taber, Mark, and I worked in the Agriculture until lunchtime, unless we were needed elsewhere that day. Laser,

Gaie, and Linda went to the Technosphere, Marine, or Wilderness biomes respectively, to continue their chores. One of us fed the animals again before lunch at twelve thirty. We theoretically took a "siesta" until two thirty, though most people worked right through, or caught up on e-mail. Afternoons were for research or more chores. Around five thirty in the evening, two of us would feed the animals. Sierra or I would milk the goats. The dinner bell rang at seven, then we collapsed, exhausted, only to get up at six thirty to do it all over again.

The weekends gave us some respite. On Saturday and Sunday mornings, I fed the animals and milked the goats, then sat around having a leisurely breakfast with the others. Afterwards we all wandered down to a large empty room next to the animal bay. Here we were supposed to practice theater for an hour or so. Often we were so tired we just lay flat on our backs the entire time.

After lunch on Saturdays, we cleaned up the week's filth. Taber and I were always on animal-bay cleanup. What a mess! I imagined with dismay the photos visitors took home of the two filthy biospherians singing along to the latest hit, spattered with pig, chicken, and goat doo-doo.

If a particular task needed attention, we would all converge on it for the last couple of hours of Saturday. Sunday afternoons were for laundry, reading, sleeping, or pursuing hobbies, such as painting or playing electric piano. The afternoon was punctuated by a very civilized cup of tea with fruit tart or some other delicacy. And on Tuesdays and Thursdays we meditated, just as we had before entering Biosphere 2. We held Tuesday soirees on a variety of topics, and philosophy nights on Thursdays. We all gave speeches on Sunday evenings after dinner.

At the weather report on our first morning in Biosphere 2, Sierra announced that the CO_2 low had been 521 ppm, up over 200 ppm from the day before. The day was cloudy, as was the next. By September 29, the lowest the CO_2 was drawn down by the plants was

826 ppm. We held an emergency meeting to discuss taking action to manage our atmosphere. We would start by preventing dead plant material from rotting, a process which releases CO_2 to the atmosphere in the normal process of decomposition. On September 30, we began harvesting the dead grass in the Savannah to halt its breakdown.

Natural savannahs have herds of bison or antelope that eat the grass. In our Savannah, humans armed with sickles were the bison. We cut clumps of dried grass, handful by handful, dropping it down an elevator shaft to the basement, where we stacked it. This was our carbon storage system, our carbon bank. We would save up carbon in the basement in the winter, then release it back into the atmosphere in the summer through composting. It was hard work, but we were resolute. No errant carbon dioxide molecule was going to prevent us from completing our two-year mission.

On that last day in September, we heard the first galago calls in the Rainforest. Beautiful black and yellow Heliconius butterflies flitted through the trees, and the chickens laid eight eggs. Linda caught and doctored a blue-tongued skink that had mysteriously sustained a bad injury. The rice had matured, and when not cutting grass in the Savannah, we waded calf-deep in cool mud, harvesting rice paddies.

There was no doubt we lived in a fish bowl. From nine every morning to five at night, there was rarely a moment when I could look up and not see tens—if not hundreds—of visitors outside, many of them pointing fingers or cameras at me or other biospherians they spotted. There was almost nowhere within the Agriculture or Wilderness biomes they could not see us. The Habitat and the basement were our hideaways, places we could get some privacy. I eventually got used to being on stage, and did not notice the eyes on me.

One morning I had to clamber into the Agriculture spaceframe to cut down vines that were blocking light. I am paralytically afraid of heights and it took all my strength to manage to creep up the spaceframe more than fifty feet in the air, to the exact place the worker had fallen to his death, a frozen smile glued on my face. A crowd of visitors had gathered below, watching my every move, waving and snapping photographs. "Get ahold of yourself," I ordered. "This is just like being up in the ratlines on the *Heraclitus*." I forced my right hand to loosen its white-knuckled grip on the spaceframe and wave woodenly at the crowd below. Camera lenses waved back.

The world outside continually interjected itself. John and Tango brought fistfuls of newspaper and magazine articles to the window, most of them positive. However, the CO_2 scrubber had caused a stir and some journalists decided that we had tried to keep it a secret. (It appeared we had.) Kathy Dyhr believes that this is what finally made much of the mass media distrust us.

As usual, I did not read the articles but overheard the other biospherians' indignation at the unfairness of the media. "The media" was a term beginning to take on the significance of a three-headed ogre.

Life inside rolled relentlessly on. Already it seemed we had too much work to do, and we held meetings to discuss how we could accomplish it all. The Savannah was still not completely cut. Peanut, sweet potato, and rice plots needed harvesting and replanting. Every day that a mature crop was not harvested and replanted was a day that new crops were not growing. This affected the amount of food on the table and the CO_2 in the atmosphere.

We all took turns cooking. It was a fun challenge to figure out what to do with a heap of sweet potatoes, a few handfuls of grain and beans, veggies of the season, and bananas or papaya. There was no rushing to the store for that special something. No opening a package. No spice shelf, soy sauce, tomato ketchup, or barbecue sauce to cover a host of cooking misdemeanors.

We had all had some practice during our Reality Experiments, but this was the real thing. There was absolutely no cheating, as there was nothing to cheat with. Burning food was a sin, which only happened a couple of times during the two years. Some biospherians were better cooks than others, but if you did not like what the cook had made—tough.

I particularly dreaded one person's cooking. His watery soups tasted like dirty grass—and quite possibly that is all he put in the pot, given how little satiation it gave. No matter how disgusting or paltry the meal was, there was no slinking off behind everyone's back to some hidden steak house, as Taber and I had done once during a particularly strenuous Reality Experiment.

Sometime during the afternoon, Sierra would leave a tub in the kitchen with all the ingredients for the next three meals. When I was on cooking duty, I always approached it with trepidation. What peculiar thing would I have to deal with this time?

The first time I cooked, I poured every ounce of creativity I had into making something that looked normal, that we would have eaten on the outside. Taber looked ecstatic when I served biospherian chiles rellenos, a green chile wrapped in pastry—not exactly New Mexico but close enough. Cheers and applause greeted my dessert of banana cream pie, the first pie baked in the Biosphere.

But only ten days into the mission, Roy thought he had broken a rib, Gaie had cut her ankle with a sickle, Sierra had injured her knee, Laser was hurting all over, and Taber's feet were so cracked he could hardly walk. We were all becoming so cranky from the change in diet that we called ourselves Grumposphere 2. Bickering over the weekly routine ensued. Some did not want to participate in theater. Others wanted informal dinners.

Then we got a wake-up call that made us all stop and take note. Life could be dangerous in Biosphere 2.

Roy and I were in the Agriculture basement, threshing rice. We were arguing about what size screen should be in the machine. He was threshing. I was cleaning rice stems from a slow-moving lower barrel that knocked the hulls from the stems. I was not paying full attention to what I was doing, and four of my fingers got caught by the barrel and sucked into the machine. Yelling and gesticulating through the noise of the thresher, I finally got Roy to turn off the machine. I pulled my hand out. As I waved it in front of me, blood splattered across Roy, and I saw the jagged end of a piece of hacked bone. I had not felt the end of my middle finger being knocked off by the thresher.

I dispatched poor Sierra and Taber to dig about in the rice to find my stump and bring it up to the clinic where Roy sewed it back on while I intermittently cursed and sang a Mexican song at the top of my lungs, "Ay, ay, ay, ay, Canta y no llores . . ." Taber helped with the procedure after he finally found the fingertip. He looked as white as a sheet of blank paper in spite of the fact—as part of his medical training to be Biosphere 2's assistant medical officer—he spent two weeks in the emergency room at the University Medical Center in Tucson. There he had witnessed much bloodier triage than this.

Later I lay on my bed trying not to move my hand. Stabbing pain pierced my painkiller haze. I felt terrible that other biospherians, already tired and overworked, had to fill in for me. Watching movies and reading to pass the time, I thought about stories I had heard of cowboys who kept right on riding after a thumb had been ripped off at the knuckle by the rope used to lasso a steer. I felt embarrassed that I was not made from such strong stock.

The media had a field day. Roy spent hours after the accident talking to reporters. People brought flowers and laid them outside the airlock. Others called with get-well wishes.

It was moving to receive such support from people I had never met. It was also bizarre. This kind of thing happens to other people, not me. At that moment, mine was the world's most famous finger.

Seventy-two hours later, Roy proclaimed my blackened fingertip dead, and sent me out to a hand surgeon to have it neatly sewn up. I could not believe it. Twelve days into the mission, after vowing not to exit for two years, I had a stupid accident that forced me to go back out.

After my surgery, a procedure that made the end of my finger look like a mini rolled roast all bound up with string, I wafted into an auditorium to field journalists' questions, my mind afloat in a sea of painkillers. I staggered out to an ambulance, which returned me to the airlock only six-and-a-half hours after exiting it. I had eaten one granola bar, drunk one glass of water, and taken approximately six thousand breaths of air from Biosphere 1.

I didn't even get a Big Mac, let alone a pizza or a margarita.

Unfortunately, as I entered the airlock, Norberto Romo, the head of Mission Control, placed a duffel bag inside the airlock with me, which I obediently took inside. I did not peek, but apparently, it only contained two computer boards, a planting plan of the Wilderness biomes, and a couple of other forgotten spare parts. This was not enough to compromise the Biosphere's material closure, but the media seized upon the surreptitious way in which it was done, causing another needless scandal. The papers erroneously claimed that I clandestinely brought supplies of food back into the Biosphere with me, and further stated that my exit and entrance negated the hermetic seal. "For want of a fingertip, might an experiment be lost?" wrote the *Arizona Republic* newspaper.

It was infuriating because they plainly did not understand that the airlock was intended precisely for such situations. A small and measured amount of air was exchanged with Biosphere 1, which would be included in any scientific calculations.

Before entering the Biosphere, I knew I would exit a different person. This accident was not exactly what I had in mind, but it was done. A small part of me would not be leaving the Biosphere at the end of two years.

I did not pull my weight for two more days, and the strain on the crew was evident. Tensions began rising. They were dispelled temporarily during Gaie's birthday party when everyone happily sipped a cup of Biosphere 2 coffee and munched birthday cake—real carrot and banana cake with banana-yogurt ice cream made by Sierra.

But our squabbling continued until even Mission Control was sucked into the emotional whirlpool, and broke a fundamental rule of keeping a team together. NASA's Mission Control absolutely does not get involved in issues that arise between crewmembers. Doing so would break down trust within the crew. And that is exactly what happened to us within the first two weeks of the mission.

Roy's personal goal was to make a documentary of our two years together. He filmed us eating, working, playing. He recorded conversations, he photographed everything.

On the second Sunday after Closure, he wanted to record our speeches. The captain, Sierra, vehemently objected to being recorded. Roy agreed, but made an ostentatious display of turning the tape recorder off when she spoke. She complained to Margret, who, instead of telling her to work it out herself, not only got involved, but punished the cocaptain, Laser, for not backing Sierra up by temporarily stripping him of his responsibility.

It was so stupid it made me want to cry. This incident and others like it sent a loud message to all the crew that we had better not talk honestly among ourselves, or we would be punished. After that, our disagreements were usually not explicit. Instead we chose to keep a lid on it, letting the pressure build internally. It was not a healthy situation.

Nonetheless, the Biosphere still needed our attention. Days were cloudy, the CO_2 was still rising, and Taber and Laser got the CO_2 scrubber up and running. We dried out the compost. In the Agriculture we replanted with sorghum seedlings instead of seed, thus decreasing the amount of time the plot would be left bare, and increasing the amount of CO_2 the plants absorbed. The CO_2 began to decline. We collectively sighed in relief.

On October 18, we broke a world record. We had been enclosed in Biosphere 2 for twenty-two days, longer than anyone had been locked inside a hermetically sealed, biological life support system. Linda had set the last record of twenty-one days in the Test Module.

Spirits were high as we drank hot tea and watched *Close Encounters of the Third Kind.* The first three weeks of our experiment had hardly been a honeymoon period. But we had reached a turning point. We were entering uncharted territory and come hell or high water, we had to explore it together.

Chapter 11

THE
LONG DARK WINTER

I dreamed of a beautiful field of purple flowers, of bean plants dripping with pods. When I awoke, I peered out my bedroom window overlooking the Agriculture, and, lo and behold, the lablab beans were in bloom, gorgeous purple blossoms covering the vines. We were running low on beans, so this was welcome news.

Because most crops take three to four months to mature, we had to start the two-year experiment with food that I had already grown in Biosphere 2. We would take this into account at the end when calculating how much of our food we were able to grow for the two years.

When we entered, we had three months of food in storage, but the beans had already become a constant problem, so we started with fewer than we wanted. The soybeans had not yielded well, and the cowpeas had succumbed to what looked like a sprinkling of white talcum powder. Called *powdery mildew,* this fungus attacks leaves until the plant simply throws in the towel and dies.

Lablab beans had saved our increasingly skinny backsides, although they would hardly be considered edible in polite society. With a long list of common names (including hyacinth bean, pig-ears, poor man's bean, *dolique d'Egypte, quiquaqua, frijol caballo, wal,* and *sem*), lablabs are generally grown in the tropics to feed animals. They were backup to the backup beans, but they were the most reliable. So we learned to cook the small black beans in the pressure cooker for an hour to get rid of toxins that made them otherwise inedible. We hid them in dishes as best we could because their flavor was dreadful, even with the added spice of hunger.

But this happy Saturday morning, October 19, 1991, I did not mourn their foul taste, but rejoiced in the prospect of soon seeing the white storage buckets filled with a continuing supply of beans, vital for our mostly vegetarian diet.

We had been hungry from the day we entered, even after meals. This was partially because, unlike most Americans who obtain far more than the recommended 30 percent of their calories from fat, we were barely getting enough at 10 percent of our caloric intake. High-fat foods are more satisfying than low fat, in part because it takes longer to digest fat than other types of macronutrients, and because fat contains more than twice as much energy per gram as carbohydrates or protein.

The other reason we were hungry was that we weren't eating much for the work we were doing. We received all the vitamins, minerals, and amino acids we needed, but we were only consuming, on average, 1,780 calories per person per day, while performing hours of demanding physical labor. We based the amount of food we ate on how much we had in storage, and what we expected to harvest over the next few months.

If we had all been desk jockeys and couch potatoes we would have had plenty to eat. The Biosphere lifestyle demanded more fuel. However, we took very seriously the goal of staying inside Biosphere 2 for two years without anything material entering or leaving the enclosure. If we had to go hungry to accomplish that aim, we would. We expected our bodies to adapt: our metabolisms would slow, and we would all lose a bit of unwelcome flab.

But after only a month inside, food had taken on great significance. We treated the first five strawberries that ripened with reverence. Laser carefully sliced them so that each biospherian got an equal share of the delicacy. When the first bunch of bananas turned yellow, we celebrated, then froze and dried those we did not eat fresh.

Our dramatic black-and-white kitchen was equipped with most modern appliances: oven, blender, microwave, refrigerator, freezer, grinder, juicer, and electric stove. Everyone put a lot of effort into cooking. We all tried to be creative with the ingredients, particularly with lunch and dinner, but no one messed with breakfast.

There was hardly a single day during the entire two years that we did not have porridge, and on the one occasion that someone tried to be creative with the gruel-like concoction by flavoring it with mint, the rest of the crew told the perpetrator to cease and desist. The message was, "You will serve porridge for breakfast, and it will be flavored with banana or ripe papaya only. Don't mess with the porridge." Everyone was very, very touchy in the morning.

Porridge gave our blood sugar level a strong boost at the start of the day. We produced a lot of flour unfit for making bread—when baked it became like a gritty adobe brick—that had to somehow find its way into the diet. Porridge was easy to make and generally very tasty, although a couple of biospherians had difficulty cooking it.

To encourage porridge-making skills we held a competition to see who could serve the best, judged primarily by taste, but texture also counted. Laser and Taber tied for first place. Each person's porridge was quite distinct. It was astonishing that so much variation could come from the same few ingredients: water, flour (generally sorghum), salt, banana or papaya. Some were more watery than others. Some were sweetened with more, or less, fruit. Roy's had a strange fermented flavor as he cooked it all night in the crock-pot, which gave Sierra a fit every time he made it. Indeed, food was extremely important to all of us.

When we entered the Biosphere, we had the idea that we would maintain, even create, a high level of civilization. But one's environment can have unexpected consequences, and hunger quickly manifested itself in some peculiar behaviors. We quickly invented the "biospherian serving," which meant heaping as much porridge or soup into a bowl as possible (the cook portioned most other food). One biospherian was particularly good at maximizing serving size to the point that a trail of overflowing soup would sometimes follow him from the serving station to the table. Not wanting to waste any food at all, we all took to sucking our chopsticks clean, and wiping our plates with our fingers. This habit soon turned into blatantly picking up the plate and licking every morsel from it. It was a great insult to the cook if all the plates were not licked entirely clean.

People's eating habits stood out, perhaps on account of cabin fever setting in. One biospherian insisted on eating with her mouth open while making disgusting noises that seemed to get louder with every passing day, though in reality I am sure that the decibel level remained unchanged. She grated on my English nerves. "We are turning into a bunch of barbarians!" I lamented, but I licked my plate along with everyone else.

CO_2 was rising again, with the daily average reaching 2,000 ppm by late October. Taber worked for several days straight, cleaning out the entire CO_2 scrubber after it clogged.

The process was horrid and he was looking distressed. Air is blown up a pipe, down which pours sodium hydroxide dissolved in water. This caustic substance scrubs the CO_2 out of the atmosphere, turning it into sodium carbonate and bicarbonate. Then, biospherians don protective gear and mix the liquid with a caustic powder, calcium hydroxide. This causes the CO_2 to precipitate out into a hard sludge of calcium carbonate, or limestone. The process then turns the sodium carbonate back into sodium hydroxide, allowing the cycle to start all over again. We were essentially performing a geologic process, only in a few days instead of millions of years.

To complete the carbon cycle, we would return the CO_2 to the atmosphere when needed by putting the limestone into a fifteen-hundred-degree oven. This was our version of a volcano. When mixed with water, the remaining material becomes calcium hydroxide again, allowing it to be used to make more limestone. It was an ingenious and efficient system but thoroughly nasty to work. Roy dubbed the thing the Hell Mines of Tarin, a reference with which I was unfamiliar but which conjured an image of a dreadful place. The protective gear of goggles, gloves, and plastic overalls worked only marginally, the fine dust particles working their way through every opening in the gear, down the neck and up the sleeves. Once, the chemicals burned Laser. I conveniently broke out in such bad hives both times I tried to shovel the calcium hydroxide that I was excused from scrubber duty. Everyone else valiantly labored on.

In our quest to keep the CO_2 under control, we planted every available spot with something green, and did whatever we could to spur on rapid plant growth. Laser and Taber built shallow ponds in the basement out of spare wood, lined with plastic sheeting. Bright lights illuminated the water's surface where *azolla,* a small water fern, grew prolifically. Mark planted hundreds of black plastic pots with sweet potato cuttings and placed them wherever they would capture some photons.

Day after day, Linda and Taber climbed into the Rainforest spaceframe to cut down the morning-glory vine, which, although green and growing, blocked light from the Rainforest. To promote new growth, we cut down every dead leaf in the banana belt around the Rainforest. (The banana belt protected the shade-loving under-story from direct sunlight.)

We laid the harvested biomass out to dry in the basement, except a few of the choicest banana and morning glory leaves that we fed to the goats. Once the material was dry, we dragged the bio-mass across the basement and down the long tunnel to the West Lung where we added it to the huge piles of dry Savannah grasses. We cut down the freshwater Marsh reeds to store the carbon in the biomass, and to stimulate the plants to grow abundant new leaves, thereby storing more carbon. Despite concern for the long-term health of the plants, Linda turned on the overhead sprinkler system to rain on the Desert to bring it out of dormancy early so the plants would start soaking up CO_2.

The lower windows throughout the Biosphere fogged up in the mornings from the cold air outside. The moisture was growing a good crop of green algae that blocked light in the Rainforest. We washed windows while a tour of visitors enjoyed pointing out spots we missed. Although all of this was extremely hard work, it felt good to be attacking the problem head on.

As we struggled to manage our atmosphere, something we are faced with today in Biosphere 1, and nurtured all the plants so they would be healthy and grow faster, I contemplated the importance of stewardship. At the time the notion was growing that we should be stewards of our earth, just as we in Biosphere 2 were stewards of our life support system, our biosphere.

However, that idea implies that we humans have dominion over the earth, and that we want it all to be our back garden, manicured or at least managed. There is no place for true, untouched wilderness in that vision, even though Biosphere 1 got on quite well without us for over four billion years.

I prefer to think that we need to be stewards of ourselves, stewards of our effects on our Biosphere, and of all human beings. We are the ones making a mess here on earth. It is we who can clean it up, and we who can help all people attain a dignified lifestyle of comfort and well being.

In a time when milk comes out of a carton and not a cow, and ecopsychologists fear that our distancing from nature is exacerbated by our first impressions being the sterile grays of a hospital and not the leafy greens of the natural world, thinking of ourselves as citizens of Biosphere 1 is a tall order indeed. Einstein said, "All my inspiration has come from observing nature, and from no other source." Somehow we have to get past nature as something we just look at. In Biosphere 2 we had no choice.

On Halloween, we even had trick-or-treaters come by the visitors' window. Frankenstein and Igor stopped by, much to all our delight. But two days later a lab technician who had been laid off played a trick of his own. He disclosed Biosphere 2 CO_2 graphs to the media and went on TV claiming that Biosphere 2 had failed and was embroiled in a massive coverup.

It was wholly unfortunate, as the data he presented were from equipment that had not yet been properly calibrated. Thus the data were inaccurate and holey, and included calibration runs, which had nothing to do with the actual CO_2 levels inside the Biosphere. The data were also taken from two weeks when storms railed around Biosphere 2, causing the CO_2 to rise dramatically. What the graphs did not show was that the CO_2 levels dipped dramatically when the sun came out. We were convinced that once sunny days returned, the CO_2 problem would resolve itself.

I wondered what had caused this disgruntled employee to act so viciously. His cries of "fake" and "coverup" were picked up all over the world, and the *Village Voice* had another go at us. Sometimes I felt like burying my head in the coral sand on the beach to block out the name-calling. Other times, I felt like walking out of the Biosphere and punching the little snot on the nose.

Shortly after this attack, a videographer, who had been given free run of the Biosphere prior to Closure to make an agreed-upon film, tried to sell it off to a third party. He was rather annoyed when SBV sued him for breach of contract and appeared on TV to throw his own mud at the project. Sometimes I thought I was living a soap opera.

As fall rolled on, the days grew shorter, the sky cloudier and the weather outside chillier. I felt out of place in my jeans and T-shirt, watching people on the outside huddled in winter woollies. The cold weather caused water to condense on the steel skin of the Habitat and it began to rain in the office, in the hallways and in our bedrooms. The Habitat was decorated with buckets and kitchen pots to catch the drips. But ours was an artificial biosphere, so we could control things like rain. Once we lowered the Habitat's humidity, the rain dried up.

However, we could not control the weather outside, and the clouds persisted. Sometimes, an interaction between the Pacific Ocean sea surface temperatures and the atmosphere trigger an El Niño-Southern Oscillation event, causing droughts in Africa, and torrential downpours on the Pacific's eastern coast. Peruvian fisherman first noticed the phenomenon, and named it "El Niño" because it would start around Christmas time. Blankets of dead fish washed ashore, killed by the change in water flows and temperature.

El Niño affects global weather patterns, and brings thick cloud cover and rain to Arizona. Rumors started that we were going to have such a winter.

The CO_2 was dancing wildly up and down depending on whether the sun shone or stormy weather darkened the Biosphere. The level had already risen over halfway toward the limit that would trigger our departure from the Biosphere. "What was the low today?" was a common question needing no further explanation. We would all study the graph, looking to see how much the high had risen or the low had fallen. In the middle of the afternoon, it was not unusual for us to start worrying where the low would be that afternoon, before the line on the graph started to rise again after the sun had dropped below the Tucson Mountains. It was nail-biting.

Gary Hudman, a member of Mission Control on the outside, worked with Linda and Taber to reduce the Biosphere's respiration rate by manipulating temperature. Colder temperatures at night would slow chemical reactions, and thus the production of CO_2. They tried to eke out every extra part per million (ppm) of CO_2 drawdown at the end of each day.

The CO_2 low for the day would occur around four thirty in the afternoon, when the sun was no longer shining directly on most of our plants. The threesome noticed that if they cooled the Wilderness and Agriculture air temperature as the CO_2 reached its lowest point of the day, the CO_2 would continue to drop for as much as 50 ppm, probably because the lowered temperature slowed plant respiration.

The rising CO_2 started causing problems in the Ocean. The pH in the water dropped below acceptable levels because so much carbon dioxide had dissolved into the water, forming carbonic acid. That would damage the Ocean's corals, which are highly sensitive to pH. The antidote was for Gaie and Taber to pour bicarbonate of soda into the seawater to raise the pH. The baking soda had been stored in the basement for precisely this situation, although we had hoped we would not need it. All in all, the two of them dumped forty-five hundred pounds of bicarbonate of soda into the Ocean over the course of the two years.

On top of the Rainforest mountain was what we called the Cloud Forest. Morning glory vines had taken it over, so we harvested them and all the carbon their biomass contained. We replanted with sugar cane, one of the fastest growing plants around, and edible to boot. While cleaning up the Cloud Forest, Linda discovered volunteer bark scorpions, *Centruroides excilicauda,* the only potentially fatal scorpion in Arizona.

We had specifically left venomous animals off the species lists, so it came as quite a surprise when, two days later, a scorpion invaded the kitchen. It was an ominous sign that called for action. This scorpion became the first species to go extinct in Biosphere 2.

When we were not struggling to keep the CO_2 in check, we were an agrarian society revolving around food. We all spent time laboring in the Agriculture, finishing the season's major harvests and replanting. The rice had produced adequate yields. We dug up the peanuts, and washed and dried them. The goats ate the protein-rich peanut greens, which they loved. Their milk production reflected such high-quality feed. The sweet potato harvest was good, which prompted a harvest festival—one entirely satisfying meal.

Soon after, we celebrated Linda's birthday, devouring the first

Biosphere 2 pizza with a sumptuous topping of chicken, goat cheese, tomatoes, and peppers. We sipped our bimonthly cups of coffee and savored the birthday cake decorated with dried fruit. Birthdays were a great excuse for celebration and feasting.

Sometimes, after dinner, Taber and I would sit on the beach and listen to the "co-kee, co-kee" of the tiny green tree frogs echoing across the Wilderness. This sound gave them their name, coquis. The ever-present chorus of crickets accompanied them, with occasional solos by our four galagos. Those monkeys vocalized in defense of their territory, which apparently did not end at the glass skin of the Biosphere. Linda reported that they were barking angrily at people outside the Biosphere, as if visitors were trespassing.

One evening Oxide, the alpha male galago, sat in the Savannah watching us all eat dinner. A couple of days later the alpha female, Topaz, attacked Opal, the beta female. They fought until poor Opal was so badly injured that Linda had to hospitalize her in a small cage in the medical facility. Despite their beguiling, cuddly looks, the nocturnal creatures have a vicious bite that can cause a nasty infection. Three times a day for ten days Linda donned heavy leather gloves to avoid being bitten and held the little animal still while I injected her with medication and doctored her wounds. Opal finally recovered and Linda released her back into the Wilderness hoping that the two females would work it out. Instead, Opal exiled herself to the basement, well away from Topaz.

On a cloudy November day, Stardust, one of the five African pigmy goats, gave birth to two healthy kids. One goat giving birth can stimulate another pregnant doe to follow. Sure enough, a few hours later, Vision, who was impossibly large around the girth, went into labor. A normal goat birth is over quickly, but after twenty minutes of hard labor only one foot was protruding and nothing more was happening. Long labor can endanger the mother's health, so Taber felt about until he found the second foot, caught behind Vision's pelvic bone. He released it, but twenty more endless minutes later, the baby had advanced no further. Taber's hand was too large to reach into the uterus. So, in consultation with Safari on the outside, who had worked with domestic animals for years, I clipped my finger nails and stuck my hand in poor Vision's vulva as she bellowed in pain. All I could feel was the tightness of her bones crushing my hand as they contracted, the kid's warm moist fur, and the uterus wall, which I had to be extremely careful not to damage.

I felt utterly helpless, having no experience in delivering a baby of any kind. None of the goats had had trouble birthing before.

Dr. Barbara Page, the vet, was on the phone with Sierra mediating. "Twist the kid, twist it clockwise," came the order.

My hand hardly fit inside and I hoped I was not poking a hole through Vision's uterus. She was crying and wriggling. I have never wished I was blind, but at that moment I did. The sensitivity in my fingers was so poor I could not make any sensible picture out of the wet warmth my fingers touched. Maybe I had the head of one kid and the feet of another.

Dr. Page stood at the animal-bay window at last, with a small kid from the Biospheric Research and Development Center in her arms to demonstrate. She had drugged the animal so it was putty in her hands. I had also sedated Vision, who was now calmer. Dr. Page walked me through the process of extracting the kid, demonstrating each move with her demo-kid. I grabbed the little lower jaw of the baby goat, which started to suck feebly on my thumb. It was still alive. Taber held the creature's legs as we twisted it into position.

By this time, Vision was exhausted and so relaxed that Mark held her head and Sierra held up her hind legs while I pulled the kid's little head through the pelvic opening. The head was so large that it barely fit. At Dr. Page's encouragement and insistence, I pulled and pulled on the poor little thing until finally it popped out and lay limp on the floor. I had undoubtedly broken its neck in the process, but we had saved the mother, who was sore, but otherwise apparently healthy. Goats usually have twins, but there was no second kid, only one enormous baby.

I checked for holes in Vision's uterus lining but found none. Time would tell, as she would be dead within seventy-two hours if I had damaged it. The following morning Vision was nursing and cleaning Stardust's kids, and three days later she was still healthy. Dr. Page congratulated us on saving the mother, and consoled me with her assessment that because of its size, I had been forced to pull so hard on the kid, thereby killing it. Life is so precious, and seemed all the more so in the confines of our little world. I was cheered to see Vision sleeping with Stardust's kids, having adopted them as her own. Stardust seemed uninterested in her kids, so she became another milking goat, boosting our milk production.

Nothing goes to waste in a small biosphere where the limits on resources are very obvious. So a couple of days later, I cooked up the kid and fed it to the two adult pigs, Quincy and Zazu. I committed a great *faux pas*, however. I placed the meal so it could not be seen from the windows, as I thought the sight of a pig chowing down on a small goat might seem ghoulish to any onlookers. Unfortunately, Quincy grabbed the little body in his mouth, and dashed with it away from Zazu and toward a crowd that had gathered, noses pressed to the windows. I ran into the pen, yelling and waving my big cooking pot around wildly in the air, chasing Quincy into the back pen as quickly as I could, hoping that the visitors were watching the mad woman waving a pot chasing a big pig, and not what was dangling from the pig's mouth.

A few days later, we narrowly averted a disaster. Taber and Linda had measured an anomalous CO_2 increase. When Mark and I walked into the West Lung late in the afternoon to move some of the hay that was stored there, we found its cause. To our shock and utter horror, puddles of condensation had collected all over the ground. The tons of biomass stored there were getting wet, which would cause them to start decomposing. The whole room reeked of mildew; fungus had already begun to grow rapidly on the grasses. If we did not do something, the rotting biomass would increase the CO_2 in the atmosphere and undo all our hard work.

The entire crew dropped everything to drag the wet material out of the lung to where it would dry out again in front of air handlers. I sewed a huge duct out of black liner material to direct hot dry air from one of the air handlers in the Agriculture basement into the lung to dry it out. We worked day and night until the job was done, all the while kicking ourselves for not having foreseen this would happen when rain fell in the Habitat.

Despite the hard work and the bleak weather, we were in high spirits as we all launched into our speeches one Sunday night toward the end of November. They were upbeat and lighthearted, some even hilarious. One in particular stands out. Mark laughed and joked his way through his speech, but his words were an ominous portent. He said that he wanted to write the story of the project because, "There is going to be a battle to define what the history of Biosphere 2 was."

At the time I thought he was being melodramatic. Looking back, I see just how prescient his utterance was. The winner of every battle gets to write its history, and Mark had foreseen the bitter battle over who would control Biosphere 2 . . . and its story.

Thanksgiving was on the way, and with it, a fresh flurry of media attention. This time the coverage was straightforward and mostly good. "What are you going to eat?" was one of the many questions repeatedly asked. The answer: everything.

That day we gorged ourselves on 3,000 calories per person, plowing through almost two days of food. We each ate a quarter chicken stuffed with ginger rice, beets in orange sauce, baked sweet potatoes, and chiles rellenos, followed by banana bread and an absurdly huge slice of sweet potato pie, a quarter of a pie each.

We washed it all down with home-brewed rice wine that was absolutely foul tasting but otherwise quite effective. It seemed fitting that we sat down to this feast just as we finished our autumn harvests. During the celebration, Mark sat cross-legged on the floor, reading a short passage about the original Thanksgiving feast. We all took turns reading from Walt Whitman's *Song of The Open Road.*

On the last day of November, I awoke in the middle of the night to an alarming crashing sound echoing across the Habitat roof. As I hurriedly turned on my light to go investigate, I glanced out into the night to see snow fluttering past. Snow had collected on the library roof, a phallic domed spire that projected above the rest of the Habitat. It was slipping off in large sheets and pounding the roof below with great metallic thuds. The next morning I felt very smug in my T-shirt, while those on the outside were huddled in their parkas. But I so wanted to go outside and make a snowman, an urge made all the worse by a snowball fight underway on the lawn outside the airlock.

The chilly, cloudy days made the Biosphere dark and dank, even in the Wilderness, which usually felt tropical. It was as if our life systems were crying out for the lost sunny days of summer. Tears dripped from the roof all over the Biosphere, even on me as I lay in bed. With the darkness, an oppressive mood hung over the crew. The exhilaration of starting the experiment had dissipated, and the tedium of daily routines and gnawing hunger were chipping away at our morale, creating a lassitude and edginess that seemed out of character.

We plodded to our daily duties, putting out little fires as they started. The rain system broke down in the Wilderness. Someone left a tap on in a bathroom and flooded the Habitat's ground floor. Fireworms were eating anemones in the Ocean, but Laser eventually caught the predators. The CO_2 was slowly creeping ever upwards, so Gaie and Laser hung lamps over the rice paddies in the basement and over the Orchard, to promote plant growth in those particularly shady spots. We lost power for nine minutes. The outage had little effect other than making us all realize how noisy the mechanical systems were. The liquid nitrogen generator—which Taber needed to operate much of the analytical equipment—began acting up.

The gray mood spread to the project management outside and the snow had apparently numbed their common sense, as evidenced by another goof with the press. For the first ten weeks of the mission, Taber and Laser had helped Freddy quantify the Biosphere's leak rate and find the remaining pinholes to bring it below the prescribed maximum allowable rate of 10 percent per year. The full weight of the lungs' metal pans rested on Biosphere 2's air volume to keep the internal pressure high. This made the job of finding leaks easier than if they had maintained a neutral pressure, which would have minimized leakage.

Taber had helped Freddy ascertain the leak rate by measuring the dilution over time of an inert gas in Biosphere 2's atmosphere. Now Freddy had found and plugged so many holes that the leak rate was down to 8 percent per year, a huge accomplishment. Freddy subsequently reduced the internal air pressure to approximately neutral by using the fan in the lungs' outer skins to suck up on the heavy metal pans to reduce their weight on the air volume.

But the process of hunting for the leaks had left us with too little air in Biosphere 2, evidenced by the lungs hanging dangerously low. To replenish the atmosphere once and for all, Freddy injected into the Biosphere a measured amount of air with known constituents that could be accounted for in future calculations and scientific work.

No press release was forthcoming, and no explanation of what had been done and why. Finally, weeks after the event and after the news had already leaked out, someone circulated a ten-line engineering specification sheet, which the journalists must have found entirely incomprehensible. Naturally, the headlines flashed that the experiment was void because air had been added, proving that the Biosphere was not materially closed as had been promised. It simply fueled the argument that the project was unscientific.

The hubbub reached the English papers, which propagated more silliness. "Big Blow for Bubble Believers," began one article, which went on to claim that we had secretly hidden a year's provisions before Closure, that I had smuggled in supplies when I went out for hand surgery, that we had fiddled the seals on the door to make it appear that we were staying inside when in fact we were going in and out at will, and that the reason the air had been pumped in was because of the rising CO_2. In short, the project was apparently a complete fraud.

My brother, Malcolm, was outraged and leapt to my defense in an interview with the paper. He was quoted as saying, in true British understatement, "If this thing is a fraud it is a very expensive fraud. I hardly think it is likely." The rebutting article was entitled, "Jane's Life in Space Bubble Is No Sham.'" It was very weird to have my life and integrity discussed in print.

I was baffled that the project management was acting so ineffectually. It seemed to me that just as the project was on the verge of being an enormous success it was beginning to falter. What went so wrong? Even now, fifteen years after the fact, I find no single clean and neat answer, but a situation full of ambiguity and paradoxes.

A straightforward reason there was such a media backlash to the project is that the media were set up. From early in the project, we presented the notion that the primary goal—and almost a *fait accompli*—of Biosphere 2's first enclosure was that it remain sealed for two years with no material entering or leaving the structure. Unfortunately, this established a succeed/fail criterion that was entirely foreign to the world of science, where success is measured by whether one learns something from an experiment. If a hypothesis is proven invalid, the experiment can still be considered a success. But at Biosphere 2, as things began to go inevitably awry in a project where something immense was being attempted for the first time ever, the cries of failure echoed.

It was not that the situation was particularly unusual. Projects running amok were certainly not confined to Biosphere 2. But we expected more from our special team (as we thought of ourselves), and from our leaders, John and Margret. What is particularly ironic is that John often accused people of "snatching defeat from the jaws of victory." I feared that we were about to do exactly that.

Dr. Richard Farson, a management consultant and Harvard psychologist who heads the Western Behavioral Sciences Institute, recently told me, "People think that organizations are strong and individuals are weak. That is wrong. Individuals are strong. Relationships are fragile. There is no relationship that cannot be broken."

Synergists were embodying these profound words, fleeing the project like the proverbial rats from a sinking ship. Some of them had been involved with John and the various Synergia projects around the world for twenty years or more. It was clear to some of us on the inside that the whole synergetic way of life was in big trouble. I hoped that the relationships among the eight of us inside could withstand the pounding they would undoubtedly sustain.

The Synergias seemed to be going down a well-documented path trodden by many tight-knit groups that follow a charismatic leader. All too often, noble ideas with a powerful visionary to guide their execution slowly become dogma and the leader becomes increasingly dictatorial. The group becomes ever more insular and the kingpin eventually despotic. It is at this point that the group either explodes by disbanding, or implodes by destroying themselves, sometimes quite literally.

The main question in some of our minds was whether the project would survive the fight.

Chapter 12

CABIN FEVER

Being enclosed began to have odd effects on our psychology. Our pasts began to tug at our consciences. Several biospherians said they felt an overwhelming urge to deal with life's loose ends. It was as if, having started life in a new world, we had to clean up our affairs in the one we had left.

One Sunday morning, I stared at the phone. Finally I picked up the receiver, and shakily dialed a number I had not called for over ten years. Suddenly, an unmistakable man's voice flooded my ears, *"Oui! Allô!"*

"Salut, Daniel, c'est ..."

"Tiens, Jane, c'est toi? Vraiment?" And I heard the gentle laugh of my first love—the ski instructor I had met in the Alps. I told him I was in the Biosphere. He told me he had two kids. I babbled about how sorry I was at how it had all ended.

He was so sweet as we chatted. He must have thought I was out of my mind. Perhaps I was.

Taber's loose ends went back even further. He called up a high-school sweetheart with whom he had had a messy falling out. This poor woman lived in Kansas with kids. They both laughed. What peculiar therapy!

The dreary weather continued and in mid-December the official word came. El Niño was here. It brought weeks of stormy days, some of the darkest in Arizona history. For ten days straight, black clouds hovered overhead without a fissure to allow sunlight to leak through. The meager light was further diminished by the glass and steel structure of the Biosphere to between 45 and 50 percent of the light outside.

The never ending darkness proved too much for some of the plants, whose defenses against disease were lowered by the reduced light. I stood helpless in the Agriculture, watching entire plots of peas succumb to a fungus that attacked their roots and quickly killed them. I knew their deaths meant less food on our plates.

Plots of white potatoes that had been lush and healthy the day before began to turn brown and shrivel. Within a week, the infected areas were dead. Sierra and I looked at the leaves with a magnifying glass and could see no insects. There was no evidence of disease on the roots, yet we could not seem to keep the plants alive.

Eventually, I looked at a leaf under a microscope and discovered a miniscule mite on the leaves. Who would have thought that such a tiny creature could have such devastating effects? After consultation with pest-control experts we concluded they were broad mites, which inject a toxin into the plant, thus causing the plants' rapid decline. These mites usually only occur on tea plants, so it was a complete surprise to find them poisoning our potato crops. We immediately decided to replace all future potato crops with sweet potatoes, which seemed immune to the pest. In the meantime we would try and save the white potatoes we had growing, although we had no experience controlling the mite, since we had never seen it before.

Everyone was becoming hungrier, although we were happy that our cholesterol levels had gone from an average of 200 prior to Closure to a low of 125 now. For the first couple of months, Roy had claimed that he had not suffered hunger pangs, but even he was now beginning to complain, even while extolling the virtues of a nutritionally complete but low-calorie diet. Linda, Gaie, and Laser were beginning to look particularly skinny, as they had come in with the least fat reserves. The rest of us still had some to lose.

Nonetheless, we were all suffering from the limited calories. We began to budget energy. We stacked items at the bottom of the stairs that we needed to take to upper floors in the Habitat. We would take them all at once, at mealtime, to reduce needless trips up the long flight of stairs. By 4:00 or 4:30 every afternoon I ran out of steam because I had used up all the calories from our lunch.

Each cook prepared dinner, then the next day's breakfast and lunch. Often not much food was left by lunchtime—extremely watery soups dominated the menu. We certainly never suffered from dehydration, but struggling with low energy made us all irritable, sometimes vitriolic.

Our life revolved around sunlight and, on December 22, we heartily celebrated the Winter Solstice, as many tribal people do. With lengthening days before us, visions of lowering CO_2 and higher crop yields cavorted before our eyes. It was a jolly occasion. The primary objective of feasts had become simple: provide more food than eight ravenous people could possibly eat.

We accomplished just that, plus a high degree of biodiversity on the serving table. We gorged ourselves on barbecued pork ribs, stuffed chickens, chile and squash, sweet corn, baked sweet potatoes, bean stew, a scrumptious salad containing many species, bread rolls, fruit cake, and Sierra's now-famous banana cheesecake made with goat cheese.

This was accompanied by another disgusting glass of *chung,* or rice wine, which we nonetheless drank as if it were a glass of the best Merlot. And of course coffee, which only appeared for celebrations. We had barely enough coffee for one cup each every two months, after someone who shall remain nameless blasted the flowers off the coffee plants in an attempt to remove pests, thereby dramatically reducing our coffee harvest.

For hours we laughed until our sides ached. Everyone told stories about what seemed to be past lives, funny events that happened in places and to people of a different epoch.

Christmas soon followed—another excuse for a feast, though not as lavish. The dining room was festive, with Christmas decorations adorning the table and room, thanks to my parents' foresight. We pulled English Christmas crackers, grimaced at the bad jokes written inside, and wore silly paper crowns that fell out of the crackers.

I had a PictureTel conference with my family in London. It was the most high-tech video link available at the time, but extremely peculiar given that the image lagged a full second or two behind the sound. Disembodied words arrived, followed by mouth movements that no longer fit the words being spoken. It was disconcerting. The image was not smooth—moving heads jerked across the screen, making me feel dazzled. But it was fabulous to be able to actually see my parents, my brothers, my nieces and nephew on the screen.

Taber popped in to say hello with a couple of friends—Spirit, one of Stardust's kids, and Mrs. Fruitcake, a hen. He was quite a hit with the children. Needless to say, several English papers wrote about the encounter.

On New Year's Eve, most of us eavesdropped by speakerphone on a Dr. John concert at the Caravan of Dreams. I felt like I was sitting right at the curving nightclub bar. Roy linked up with people all around the world, via the now-defunct technology of videophones. People on either end of the phone line had a six-inch screen with a built-in camera which took very low resolution black and white pictures, and at the push of a button, sent the image down the phone line, cutting off all conversation for several seconds as the photo transmitted. Line by line the image would appear on the recipient's screen, and then conversation could begin anew. It was tantalizing to see the smiling faces of known and unknown people slowly emerge. We felt connected to the outside world in ways that were otherwise impossible. The technology was so primitive, but we loved those videophones, and made art out of each image we sent.

The lines of communication we had were tentacles protruding through the walls around us to the world outside, feeling around in the dark to snatch up little pieces of information about the universe. In everyday life, we have so many tentacles that we fight to filter out the information overload.

In Biosphere 2, we actively sought to reduce the filter our walls threw up. NASA later learned from astronauts living on the Russian Mir Space Station how important electronic communications are to those in isolation. I find it hard to imagine how difficult it would have been to be completely sealed away from the rest of the world, like the crews of old ocean-crossing square riggers. Not only did the encouragement from loved ones boost us; so did our ability to communicate with someone other than our seven fellow inmates. These interactions reminded us that there was a whole big world outside. If communication had been cut off, we would have become even more inward-looking, even less able to keep things in perspective.

All during the winter holidays we continued the daily grind: planting wheat, butchering Stardust's two kids, culling the Asian jungle fowl chickens, weeding the Desert. When Taber was not tinkering with the nitrogen generator to keep it going, he was babying the CO_2 scrubber. Gaie had begun weeding unwanted algae from the coral reef, usually eaten by an abundance of grazing fish in a natural reef. The fish populations were not yet large enough in our artificial reef. Linda was constantly monitoring, weeding, and pruning the Rainforest, Savannah, and Desert. Mark was still planting pots of sweet potatoes and lablabs that he dotted all over the Biosphere in every spare moment.

Snow fell again, heavy and thick, blanketing the Biosphere and our spirits. The mad media circus was beginning to get under everyone's skin. It may have been ridiculous to get so worked up over what a few unqualified "experts" were saying about us. But, the misinformation was unnecessary and upset us all.

The battle over Biosphere 2's credibility was being fought in the most public arena: on television and radio, and in newspapers and magazines. Roy was particularly concerned about it and called a meeting to discuss what the eight of us could do from inside the Biosphere. Mark recalls, "The posture had been, 'we need to protect our technology, as we had put 150 million into it.' There was a culture of, 'we're a private venture; we don't need to be telling anybody anything.' We were going to be slaughtered [in the press] unless we absolutely were transparent and gave them everything."

Roy wrote a memo to Margret with our thoughts and concerns. She responded that two new people had been hired to run the PR department and would be arriving within the next few weeks. We hoped they could stop the runaway train.

Despite the continuing bad press, the muckraking, and the bad-mouthing, the public still seemed to be behind us—or was at least enjoying the fight. During an interview, *Good Morning America* cut Roy off midsentence for a commercial break. Angry calls inundated the place, demanding that Roy be given time to tell what was really going on inside Biosphere 2.

Visitors kept pouring in to see what all the fuss was about and catch a glimpse of us monkeys in the cage. Thousands walked outside the Biosphere daily, waving, smiling, pointing, and clicking cameras at us. I did not mind being watched. After all, these people were helping pay to keep the mission going. I hoped that they would gain some insights from touring around our tiny world.

However, when people banged on the glass, it made me mad. It was rude. It was invasive. I found it particularly irksome on Sunday afternoons when I sat on the balcony overlooking the Agriculture enjoying a quiet cup of tea with friends, only to be interrupted by eager photographers who, after rapping on the glass to gain our attention, would begin clicking and flashing. I will never, ever, knock on an aquarium again, or rattle bars at the zoo.

Unfortunately, all the strain of limited calories, hard work, being locked up, and the disagreements with the central management team outside began to stress our relationships inside. Rumors ran through the crew that Roy was going to quit; Linda was going to the press or Ed to get Margret fired.

Neither rumor held any water, but over long, difficult discussions, it became clear that several biospherians were very unhappy with our social structure. We had entered the Biosphere believing that it would be operated as a ship, with mates reporting to a captain. The heads of each area, which included all the biospherians, were the mates, making everyone essentially equal, although the captain would have the last word. Meetings would be run using Robert's Rules of Order with the captain chairing. This had already transformed itself into a strict hierarchy, with some biospherians distinctly lower on the ladder than others.

We had expected to help make major decisions that affected our life support system. But that was increasingly not the case, as evidenced by the breakfast meeting where Gaie ordered that the CO_2 scrubber be turned on—no discussion. This was hard for some of us more academic or anarchistic types to stomach. Moreover, none of us had ever suffered from a diminished ego.

Our discussion turned to mistrust and how difficult it was to be honest among ourselves for fear of retribution from Mission Control. We bemoaned being on stage so much. People got heavy loads off their chests. The air was cleared and we went on about our business of caring for our Biosphere.

In mid-January, sunny days blessed Biosphere 2 and the CO_2 began to plummet. We turned off the CO_2 scrubber, which heartened Taber considerably. He no longer had to spend half of every day in the dungeon with the machine. The CO_2 was going down so rapidly that we were now concerned that it might bottom out without leaving enough carbon in the atmosphere for the plants to grow and produce food. So Taber trudged back to the dungeon. With Laser's help, he prepared to release CO_2 in the scrubber's limestone back to the atmosphere.

The highest level the CO_2 had reached before the sun came out was approximately 3,600 ppm, nothing to get too worked up about—it's not until the CO_2 reaches 8,000 ppm that some get headaches. It appeared that our hard work had paid off and the worst of the CO_2 battle was over. The following winter would not be as difficult—NASA scientists studying the Earth's climate assured us that El Niño was not known to occur two years back-to-back.

Some people suffered tremendously from the hunger. Sierra agreed to double the peanut ration, and increase the amount of frozen banana and beets in our diet. The next two months would be the worst, and then vegetable and fruit production would begin to soar.

In the meantime, we continued battling the broad mites that were attacking the white potatoes. We discovered that their eggs die at relative humidity levels below 30 percent. I worked with the Energy Center and Nerve Center teams to get the Agriculture humidity as low as possible. Sierra and I sprayed the potato crops with soapy water in hopes of suffocating the little buggers.

I even tried hair dryers to kill the mites, which seemed to work by desiccating them. Teams of biospherians waded through the plants, hair dryers in hand, fighting the war on broad mites. It would have been hilarious if it were not so serious. We were battling the bastards for our lunch. Eventually they won. Every crop of white potatoes blackened and shriveled.

Because we were so tired, we stopped turning up for weekend theater. But we still had some energy for artistic endeavors. Even the most destitute people on earth have their art. For many it is sacred dances, their own bodies and music. For Taber and me it was music. We had brought a whole array of musical instruments in with us, including a synthesizer, sampler, Midi, and computer recording system. We wanted to manipulate sounds from inside Biosphere 2 and incorporate them into expressive music.

Our music making was a way to explore the difficult emotions we were feeling while trapped inside our very artificial situation, to express ourselves freely and transform those feelings into something beautiful, something moving. That the end result would hardly land us a contract with a record label did not matter. It was the process that was cathartic. We had finally finished our first piece, to which we subjected the rest of the crew, who were polite. Some were even enthusiastic.

Almost every biospherian had a background in some form of art. Mark was writing poetry, Sierra played her flute, Laser ran around filming. Roy was participating in a worldwide performance-art piece with a close friend and actress in L.A., Barbara Smith. In a role reversal from Homer's *Odyssey,* she would perform around the world, he via videophone from the confines of Biosphere 2.

Art was a very important part of our lifestyle. Art dragged our weary souls out of the details of our busy lives, to a different plane. It would have been all too easy to sink deeper and deeper into the daily tedium and the problems at hand, but art makes us raise our heads, look up, and see that there are other possibilities, other ways of seeing the world and being in the world.

So, while our table manners had become barbaric, at least we could find some sense of civility and celebration in our art. Even cooking became an art form. Everyone tried hard to be creative. Sierra made the most spectacular birthday cakes, decorated luxuriantly with fruit. Taber concocted "dinosaurs," inventively-shaped pastry shells with a bean and sweet-potato filling, or sometimes fruit. Roy concocted a papaya salsa. Laser's crepes were to die for.

We also enjoyed less high-minded forms of entertainment. We all had televisions in our rooms (except Linda, who did not want one). Taber and I started watching so much TV in the evenings that we banished the idiot boxes. On Saturdays, the person on watch in Mission Control "piped in," as we called it, a movie via the cable network.

Pascal Maslin, who was training for a later mission, was particularly dedicated to fulfilling our video viewing pleasure. We were always three months behind on new movies, and spent the first few months catching up on those we had not seen while all our attention was focused on building Biosphere 2. Sci-fi was a definite favorite and *Star Trek* was a must-see.

In early February came a flurry of activity as Mission Control and some of the crew prepared for the second Science Advisory Committee review meeting to be held on February 10. The first had apparently been held only a couple of days prior to Closure, though I had been so focused on getting ready to walk into the Biosphere that I had been completely oblivious to it.

The committee was intended to provide scientific guidance and credibility to the project. It included Dr. Thomas Lovejoy, a pioneer in conservation biology in the tropics, and at that time, the assistant secretary for external affairs for the Smithsonian Institution, as chairman; Dr. Gerald Soffen, who had led the Viking Mission to Mars and was director of university programs for NASA Goddard Space Flight Center; Dr. Stephen O'Brien, a renowned geneticist and head of the Laboratory of Viral Carcinogenesis at the National Cancer Research Institute; Professor Eugene Odum, one of the fathers of systems ecology and director of the Institute of Ecology at the University of Georgia; Dr. Ghillian Prance, director of the New York Botanical Gardens, and later Kew Gardens in England; Dr. Robert Peters, director of the Global Change Program for Conservation International; Professor Keith Runcorn, an English geophysicist who was a key contributor to proving continental drift and the theory of plate tectonics; Robert Walsh, vice president of research and development for Allegheny-Ludlum Steel.

Much to everyone's relief, this second meeting seemed to go off without much trouble. The members of the Science Advisory Committee, or SAC, complimented the team on our efforts, were concerned about some of our gaunt looks, and expected to see a full research plan for the remainder of the two-year experiment by the next meeting in July. It was a positive meeting with nothing unexpected. I greatly valued their support, particularly in such difficult times.

Another high note came with a conference that NASA invited us to attend via video. Each biospherian gave a talk on our experiment thus far and its applications for space bases. It was thrilling to know that many of the people who could really make or break the project were still behind us, admiring our efforts, despite the Media Ogre's bellowings.

Spring was on the way and the galagos produced a baby. It was healthy and the mother was taking good care of it, a triumph since galagos rarely bear offspring in captivity, and the mothers often reject the babies. It was all the more poignant as the adult monkeys had lived in tiny laboratory cages before moving into Biosphere 2, which by comparison was a paradise.

March 13 was my thirtieth birthday. We celebrated with a formal dinner set up on the beach, with everyone clad in their best bib and tucker. Taber had fashioned a chandelier out of our Christmas lights, which illuminated the feast. Linda made a superb lasagna, biospherian style, and Sierra produced one of her gargantuan cakes. It was an evening filled with nostalgia and yarns of the good ol' days of adventure on the *Heraclitus* or in the outback.

Everyone had stories. "What about the time we dived on Angarosh, Mother of Sharks?" asked Taber positively piratical as he acted out the story. He told of four of us, including Laser, himself and me, diving into waters teeming with fish and baby sharks, neon-blue pilot fish leading them by the nose. As the four of us swam further windward, we saw the huge shadows of adult sharks lurking like hungry ghosts in the distance.

"We had no weapons to defend ourselves," he continued, "Except Laser had a broomstick." There was a peal of laughter from the dinner companions.

Laser jumped in, "Hey, shark eyesight's terrible. They only see shapes, so if I hit the shark on the nose with something thin that it can't see." He picked up his chopstick and started pounding on a jug in the middle of the table. "It gets frightened and swims away. . . . Or, at least, that's what we were told by people who feed sharks for a living."

Taber continued, almost whispering, and everyone leaned forward to hear. "The sharks were large now, about twelve feet, and close. We were about to turn back when one of them swam toward us, swimming back and forth. I could see its teeth. We were backed up against the atoll. The shark swam right up to us. It was within three feet of us now. I could have reached out and touched it. There was nowhere to hide, except behind Laser's broomstick."

"What did you do?" cried Sierra, excitedly.

"We all hung there, frozen, hardly breathing. I could feel every hair follicle on my body. I talked to myself. I sensed myself. I counted its teeth. I knew that if any one of us panicked it would be the end of us all, as we were within smelling distance of thousands of sharks." He paused, and shrugged his shoulders, "Then the shark swam away."

Gaie took us on a magical tour through the Antarctic in the *Heraclitus,* finding the way through fields of icebergs at night and in fog in a ferro-cement ship. She described the creaking sound of the ice, and how she collected blubber samples from whales feeding there by harpooning them from the zodiac with a special collection dart.

With full bellies and full hearts we left dinner to go to bed. We had had a lot of good times together. I loved the way of life the Synergias offered. Where else could I be a cowgirl in the outback, sail around the world, live inside the Biosphere, discuss philosophy while cleaning windows in an artificial rainforest? But we all felt the dangerous undercurrent building.

I went to bed exhausted, sad, and confused. I prayed that the longest, darkest winter I had ever known would soon be over.

People have often asked if I suffered claustrophobia while living inside Biosphere 2. Not at all—as this photo of the one-million-gallon ocean shows, it's spacious. I often spent time on the beach—seen in the distance—to be by myself. *Courtesy of the Roy L. Walford Living Trust.*

Here the Savannah is in full growth, taking CO_2 out of the air and pumping oxygen into it. Once Linda turns off the overhead irrigation, the Savannah goes dormant and has little effect on the atmosphere. Because we could turn the Savannah on by raining, or turn it off by drying it out, we dubbed it the Biovalve. *Courtesy of the Roy L. Walford Living Trust.*

One of the four galagoes that roamed the Wilderness biomes at night. *Courtesy of Terrell Lamb.*

I'm harvesting sweet potatoes. We ate so many that our skin turned orange. *Courtesy of the Roy L. Walford Living Trust.*

Roy in the only area in the Biosphere that was locked—the banana room. The sweet fruit was far too tempting. *Courtesy of the Roy L. Walford Living Trust.*

Taber and I are butchering two goat kids—a skill we both learned during training in the Australian Outback. *Courtesy of Terrell Lamb.*

Sierra and Mark prepare for a feast in our kitchen, where we had most modern appliances (electric stove, refrigerator, microwave, blender). On the counter are radishes, zucchini, acorn squash, green papaya, jalapeños, lettuce, flying saucer squash, and bananas. *Courtesy of the Roy L. Walford Living Trust.*

Although we were hungry almost all the time, our monthly feasts were a time to stuff ourselves. This is our first gargantuan Thanksgiving meal. In the foreground sit banana bread, sweet potato pie, and cheese cake. Behind that are a salad, rice with peanuts, pork, several veggie dishes, salsa, soup, the results of one of our brewing attempts, and doughnuts behind the carafes of mint tea. *Courtesy of the Roy L. Walford Living Trust.*

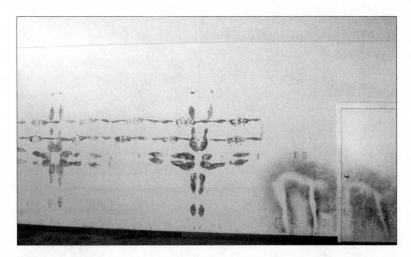

Roy's Butt-Wheeled Wagon project. We all got naked, painted body parts and pressed them against the white wall. The spokes of the rear wheel were men's butts, the front ones were women's behinds, the wagon was forged from men and women's joined hands, and Taber and I pulled the cart. *Courtesy of the Roy L. Walford Living Trust.*

Margret Augustine, the project's CEO, and Phil Hawes (TC), Biosphere 2's chief architect. *Photo by Peter Menzel.*

Workers bolted together thousands of steel struts to form the Biosphere's skeleton. Here, builders are riding a section previously assembled on the ground to its resting place on the rainforest. Many members of the construction team were mountain climbers as much of the work entailed hanging from ropes at high elevations. *Courtesy of the Roy L. Walford Living Trust.*

Biosphere 2 covers 3.15 acres. It contains five wilderness biomes (Rainforest, Savannah, Desert, Ocean, and Marsh) and two anthropogenic biomes (Intensive Agriculture Biome and Human Habitat). Each crewmember had a private sleeping area with a private bathroom shared with one other biospherian. These were above the area in the Habitat where the domestic animals (goats, chickens, and pigs) lived. The kitchen, dining room, clinic, analytical laboratory, computer room, machine shop, and exercise room (which we rarely used) were also housed in the Habitat. *Courtesy of Terrell Lamb.*

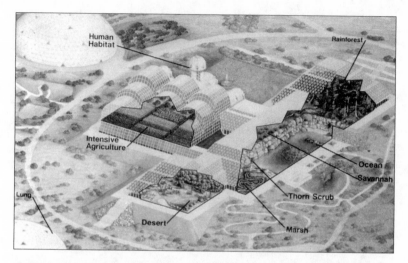

The 3,800 species of plants and animals that lived in Biosphere 2 came from around the world. We collected many of the rainforest plants in Venezuela and Puerto Rico. The Savannah plants came from French Guyana and Australia. Baja California provided many of the Desert species. We gathered corals for the Ocean in the Bahamas and the Yucatan Peninsula in Mexico. The mangroves for the Marsh came from an area in Florida that was destined to be a parking lot. *Courtesy of Terrell Lamb.*

John Allen as Vertebrate X inside the Test Module, measuring his blood-oxygen level with an oxymeter. This was the first time anyone had sealed themselves in a chamber with plants in soil, and doctors were concerned John might succumb to a lung infection or other horrible demise from toxins in the air. After 72 hours he emerged jubilant and healthy. *Courtesy of Terrell Lamb.*

We made it! On September 26, 1993, eight healthy biospherians re-entered Biosphere I after two years and twenty minutes inside the hermetically sealed structure. *Courtesy of Terrell Lamb.*

Press crews outside the Test Module await Vertebrate Y (aka Abigail Alling) at the end of her five-day stay in the sealed chamber. *Photo by Peter Menzel.*

Chapter 13

IT TAKES FOUR MONTHS
TO MAKE A PIZZA

"Break time!" I yelled, the sound echoing across to the others stooping in the rice paddies. Everyone straightened up in unison and began to wade to the edge of the fields, like a flock of herons stalking through the rice in search of food. After hosing the dark gray mud from our calves and feet we all trudged out to the Plaza for our morning peanuts and mint tea.

It was springtime. The sun was out and this morning every biospherian was working feverishly to save our young rice crop from the ravages of an infestation of loupers, moth larvae that can eat every leaf off the plants in only a few days. I had already sprayed a natural insecticide on them, but it was not working and the green half-inch-long caterpillars were growing rapidly and beginning to wreak havoc. As the bacterial killer had been ineffective in halting the tiny green army's rampage, we were picking each insect off one by one. I fed them to the chickens, who did not seem impressed.

As part of our pest management program, I had released seething bagfuls of ladybird beetles (also called *ladybugs*) prior to Closure in the hopes that a healthy population would thrive in the Agriculture. These little red-and-black bugs are highly effective carnivores. I also let loose predatory mites, green lacewings, and other bugs. Millimeter-long parasitic wasps were commissioned to attack aphids. Once an aphid infestation broke out on a plant, aphid-sized brown balloons quickly sprouted up all over it. This meant the female wasps had laid their eggs inside the aphids, transforming the herbivores' bodies into a home and pantry for their offspring. The insect world can be gruesome when viewed through a magnifying lens.

We also washed off pests, pruned damaged and infected leaves, and sprayed soap on them to suffocate insects. We rotated crops in fields to keep root diseases to a minimum. Perhaps most important, we had planted varieties of crops with specific disease resistances. When all else failed, we embarked on a search and destroy mission, picking off every insect as we were doing today in the rice.

In many ways we had stepped back in time, rejecting the yield-maximizing techniques of the Green Revolution, which required petrochemical fertilizers, pesticides, and herbicides. We could not use them because some are horribly toxic, and because we would not be able to make them inside the Biosphere, and would eventually have to import more, which went against the notion of our little world being self-sustaining. Instead, we fell back on the ancient techniques of organic farming that people practiced thousands of years ago in China and elsewhere.

Like the ancients, we understood the importance of complete, unending cycles such as the sun's as it blazed across the sky. Solstices and equinoxes were our major holidays. Chinese farmers particularly understood the need for complete nutrient cycles, returning human "night soil" to the fields. Our and our animal's waste returned to our fields via the marsh lagoon waste-recycling system and the compost machine, a grandiose term for a large hammer mill atop a holding tank.

With the lengthening sunny days of spring, the carbon dioxide level in the Biosphere was rapidly declining, so we started up the compost machine. Laser was the compost king, and most mornings for the past week he had been feeding crop residues and animal waste through the mill, situated on a platform at the top of the scarlet staircase in the center of the Agriculture. Ground-up waste

dropped into a six-foot diameter, ten-foot high barrel in the basement below to begin decomposing. The barrel originally housed an
auger that turned the compost occasionally, then spewed it out a
hatch on the front of the machine once it was ready to use. But the
auger invariably jammed, so we removed it. Instead, one or more of
us clambered inside the barrel to turn the compost or shovel out
the finished product. The compost "machine" was now merely a
compost heap. But it maintained our soil fertility as it has around
the world for millennia.

Also like Chinese farmers, we divided our family-sized half-acre
farm into small plots, eighteen in all, so we could rotate a wide
variety of crops all year, according to the season and what we
needed: more grain, more beans, more starch. One area—Sierra's
kingdom—grew only vegetables. She produced lettuce, green beans,
bell peppers, carrots, cabbage, cucumbers, eggplants, kale, onion,
pak choi, snow peas, squash, chard, and tomatoes, from a nine-
hundred-square-foot area.

Our vegetable patch reminded me of the allotments back in
England. Dating back to the eighteenth century, families were
entitled to a small parcel of government-owned land to grow fruits
and vegetables, particularly during wartime to flesh out the
rations. Like the rambling allotments, Sierra's vegetable patch
was a treasure trove.

I modeled our Agriculture on Chinese farming, and my bookshelf held several tomes dedicated to its history and techniques.
The Chinese long ago devised a synergistic rice paddy system similar to the one we were lovingly protecting from the little green
army of caterpillars. It is a brilliant way to maximize resources.

Swimming around the plants were tilapia fish, which live largely
on small creatures growing in the paddies that would otherwise
attack the rice. The fish also grazed azolla, a small fern floating
thickly on the water. The tilapia nosed around in the roots for food,
thereby aerating the plants with oxygen, and they added nutrients

to the water via their feces, thereby boosting rice production. When we harvested the rice, we also harvested fish dinners!

The rice paddy-fish system used the natural structure of ecosystems to increase farmland productivity, as did the tropical orchard. Banana, papaya, and guava formed the upper canopy, and shade-loving (and disgusting) taro absorbed the remaining photons below. Sometimes chickens roamed the orchard, eating insects living in the leaf litter. Organic farmers from many cultures and ages have viewed—and still do view—their farms as ecosystems. So did we.

From the viewpoint of a NASA life support system, the Biosphere 2 Intensive Agriculture Biome was a closed-loop, bio-regenerative, nonpolluting, self-sustaining, intensive agriculture system. That is to say, while energy would be allowed to enter from outside, no material needed to be added to the system for it to function (closed loop); everything recycled biologically instead of through physical or chemical means (bio-regenerative); it did not result in any toxic byproducts (nonpolluting). The farm would continue to be productive for years using only the materials available from within the Agriculture (self-sustaining); and it produced high yields with diverse crops (intensive).

This is the type of system that will be required for permanent bases in far away places such as Mars. Shipping costs to Mars would be outrageously expensive, and the turnaround time on an order would be at least nine months from the time the requisition was received on Earth, so a base would need to be independent.

NASA's systems are as diametrically opposed to our Agriculture as they could be, although they fit almost the same definition as our farm. Where I considered a diverse population of bacteria a vital part of a healthy agricultural soil, NASA scientists in those days sterilized every part of the growing system. The word *bacteria* was synonymous with *pathogen*.

Today, NASA experience has shown that a robust ecosystem of many different species of bacteria actually resists infection better than attempting pure sterilization. If a pathogen enters an empty system, one where no other bacteria live, the pathogen can run rampant with nothing to check it other than chemical or environmental control. But, if the same pathogen enters a thriving ecosystem of bacteria, they go to war to exclude the interloper.

Prudence would have dictated that we have a means of sterilizing the soil during our mission just in case pathogens overran it, but John had roundly rejected this notion. Good was going to have to triumph over evil in our farm soil.

So here we were, in the world's first enclosed, artificial, state-of-the-art biosphere, up to our knees in rich mud, doing what agrarian communities have done for millennia—protecting our crops from attacking insects. I could not help but chuckle at the irony. I reveled in the notion that people on Mars could be doing a similar thing one day, although we should be able to preclude most pests from a Martian farm.

After our fifteen-minute break, which we took every weekday at 10:30 a.m. sharp, I went into the animal bay to check on Sheena, the African pigmy goat. She was very broad across the beam and expecting any day now. As I entered the horseshoe-shaped room my heart leaped and I grabbed my two-way radio.

"This is an announcement to all biospherians. Sheena has had three healthy kids, over!"

"Three, over?" Sierra was appropriately surprised—goats usually only have two. But Sheena, our prize producer in every way, had outdone herself. It meant more milk for us in a few weeks, and meat shortly thereafter.

When in the animal bay, I often thought about how we received all this wonderful milk, eggs, and meat essentially for free. The miniature chickens, pigs, and goats lived off stuff we could not eat. When we harvested a plot of peanuts, we would get the nuts, and the four does would transform the greens into an average of five ounces of milk per person per day. They particularly liked bean pods, which we fed as a treat to keep them busy while we milked them. Biosynergy was working for us again.

The next day Chris Helms, now head of the project's PR, visited and gleefully displayed the front page of a local paper with a photograph taken through the glass. It showed me cuddling the triplets with the headline "New Kids on the Block."

The kids were utterly beguiling! After only a few hours they were skipping about the pen, happy to be alive. I always named the kids. I espoused the old adage that a well-loved animal makes the best meat and Sheena's little black kid became "Boing-boing," as its legs seemed to be made of springs.

The animals were good for the soul—most of the time.

One morning I walked into the chicken pen, and picked up a hen that had not laid any eggs; we could not afford slackers. I carried the bird into the small butchering room and gently placed her on the cutting board, stroking her long neck to soothe her. I hated the next bit. But Roy was filming so I could not hesitate.

I took a sharp knife, laid it across the animal's neck, took a deep breath, and sliced off her head. In an instant, the bird leaped up and off the table, and ran out of the butchering room. I shrieked with surprise, charging around the animal bay trying to catch the headless chicken. I finally grabbed the hen and dumped her in a pot of boiling water to loosen her feathers for plucking. It was like a Quentin Tarantino movie, blood squirting from the neck stump. Roy chuckled at the black humor—he had captured the scene for posterity.

I was finding it harder and harder to butcher the animals, though. Living on a mostly vegetarian diet, I had lost the enzymes that help digest meat. So I was giving all mine to Taber. I felt even more connected to the darlings once I knew I would not be eating them.

A few weeks earlier, Zazu had given birth to seven healthy piglets and three chicks had hatched. The animal bay was crammed with little animals squealing, bleating, snorting, and chirping. The big bay—about thirty feet by fifty feet—had large windows around most of the twenty-foot high walls. The floor of two of the five pens was a one-foot deep soil bed for chickens to scratch and peck. Here I collected the few eggs that the small Jungle Fowl hens laid. The other pens, where goats, chickens, and pigs ran together, had concrete floors that were easy to clean.

The meat from the livestock provided an insignificant average of 43 calories per biospherian per day, all eaten at Sunday dinner. The milk provided only 100 calories per person, but 16 percent of our badly needed fat intake. Peanuts provided another 36 percent, and bananas a surprising 31 percent.

Life in Biosphere 2 was often reduced to numbers.

- 257, 898, and 134: the number of eggs, the pounds of milk, and pounds of meat produced in the first year.
- 30,000: the approximate pounds of soil in Biosphere 2.
- 1,000: the gallons of water condensed out of the air every day for showers, drinking, and irrigation.

Bananas turned out to be one of the most important ingredients in our diet. In the first of our two years in Biosphere 2 we ate just over a ton of bananas, only slightly less than the 1.3 tons of sweet potatoes. The 208 calories a day we each ate were manna from heaven and I still eat a banana a day. Bananas were the sweeteners in our desserts and ice cream, and they were the thickeners in our pies and puddings (bananas contain a substance that is similar to pectin which thickens upon cooking). We froze them, we dried them. They even made the best booze, although we did not ferment many, as it was a waste of good food. The banana storage room was the only locked room in the entire Biosphere, as the yellow fruit hanging in huge bunches from stainless steel chains was too much of a temptation even for highly disciplined biospherians.

Since the white potatoes had succumbed to broad mite, sweet potatoes had taken their place in our diet, providing over half our daily calories. We were eating so many that we were turning orange from the beta-carotene. Friends and colleagues invariably noticed the bright orange calluses on our hands when we put them up to the window for our biospherian handshake.

I still eat the occasional sweet potato, but I do not care if I never see another beet again as long as I live. Beets grew prolifically, and we left them in the ground longer than normal, greedy for every extra bit. The result was woody, tasteless borscht, day after day. But as my mother would say, "They kept the wolf from the door," particularly in the winter months, when few other vegetables grew. The first time we ate an abundance I thought my bladder was bleeding—it turned out the beets had colored my urine pink.

Since we were mostly vegetarian, beans were vital to our diet. Soy and peas provided some variety, but the nasty-tasting lablab beans continued to be the most prolific, winding high up into the spaceframe and flopping over the balcony to form an edible curtain where they entwined the leaves of the papaya trees, our other prolific fruit producer.

Papayas were a biospherian delicacy. I loved to slice open the beautiful pumpkin-orange fruit and clean out the glistening round black seeds. The tiny marbles were so soft, so perfectly round, such a wonderfully contrasting color to the orange flesh. The ripe fruit's fragrance was heavenly. Along with the bananas, the papayas made our morning porridge tasty.

For grains, we grew rice, sorghum, and wheat. The wheat variety, Yecora Rojo, was developed for and grown in space by scientists at Utah State University. They gave it to us to grow since it was a high-yielding, short plant. It wasted little energy on growing stems and leaves. In March, we harvested the first wheat fields.

It was a decent harvest, so I made a pizza to celebrate. As I savored it, I considered how it had taken four whole months to make. We had all participated in the creation of this pizza, either by planting, watering, weeding, or finally, harvesting. Never had my connection to my food been so direct or so satisfying. Here is the recipe for our pizza.

TOMATO AND GOAT CHEESE BIOSPHERIAN PIZZA

The Flour

Rake and level one field.

Drop one wheat seed every six inches into four-inch-deep furrows made with a hoe.

Cover with soil, water, and let sit for three months while the plants grow. Intermittently weed, irrigate, and control pests so the plants produce healthy heads of seed.

After three months, or when the plants are beginning to brown on the edges, turn off all irrigation and let sit another few weeks.

Once the plants are golden brown, harvest with pruning shears (do not use a hand scythe as this shakes loose precious wheat grains, wasting them on the ground).

Process all the wheat in the threshing machine, being careful not to get your hand caught. Sift out remaining chaff. Place wheat stalks in the animal bay for bedding.

Set aside some grain to replant wheat crop.

Grind half a pound of kernels into flour and carry up to the kitchen.

The Topping

Feed four African Pigmy goat does banana leaves, elephant grass, and crop residue. When each is in heat, place with Buffalo Bill, the buck. Wait until babies are born.

Help mother with birth if necessary and wait another six weeks to wean the kids.

Milk does.

Set milk on back kitchen shelf for two weeks to ferment. Once thickened, set in cheesecloth and hang twenty-four hours to strain. Carry into the kitchen.

The day of making the pizza, collect one pound of tomatoes, two Anaheim chiles, three jalapeños, and herbs as available.

The Pizza

Take the half-pound of wheat flour and mix with two tablespoons of live sourdough that Sierra continually maintains. Mix with water until a doughy consistency, knead and let sit overnight.

Roll out the dough into two nine-inch pie pans. Prick with a fork.

Cook two sweet potatoes in their skins.

In the blender, puree tomatoes with sweet potatoes as thickener and season to taste with sea salt stolen from the Ocean supplies. Spread evenly over piecrust.

Sprinkle with cheese, sliced peppers, and herbs.

Place in oven at 350 degrees and cook until cheese has melted and the crust has turned a golden brown.

Remove from oven, slice and serve piping hot.

Although I had projected being able to grow only about 80 percent of our food, we were in fact doing much better than that. We were eating entirely from what we were growing—100 percent Biosphere 2-grown food. Unfortunately, we accomplished this by eating a great deal less than we would have liked. We were all losing weight. The guys had lost on average 18 percent of their weight. Taber had shed almost sixty pounds, and was so skinny that I started giving him some of my food. The women lost about 10 percent.

None of us was in danger of malnutrition, as the diet was complete. Roy called the diet nutritionally "superb." However, the protein content was marginal for highly active people at an average of 63 grams per person per day. By contrast, body builders eat 1 gram of protein a day for every pound of body weight. For building muscle rapidly, I should have eaten twice the daily amount available, as I weighed under 120 pounds.

However, we were not trying to become Mr. and Ms. Universes. After testing our urine, Roy determined that we were just barely getting enough protein. Our diet was simply low on calories, so our bodies were eating themselves, taking all the fat reserves and then tucking into muscle. We were becoming svelte . . . but weaker.

We were far from starving, but I was beginning to understand the terrible plight of people in the world who are truly hungry, fighting for their lives and for their family's lives. Aside from the misery of hunger itself, it is a dreadfully helpless feeling to not have enough energy to fix the problem that caused the hunger in the first place.

Farming is bloody hard work.

Long before the next meal arrived, I had usually burned up the calories from the last one. I dragged myself from chore to chore, taking twice as long to weed a sweet-potato field as it should have taken, skipping other chores. I am used to being strong and vital and I felt far from that most of the time.

It is impossible to say whether the bone-deep weariness most of us felt reduced our yields, but I am certain it left little energy to find creative ways of growing more food.

Roy was having a whale of a time, however, because this was his life's work. Before entering Biosphere 2, Roy had been studying the effects of our type of diet on mice. He found that a low-calorie, nutrient-dense diet makes mice live to the equivalent of 120 in human years. That does not mean we humans could have 40 more years of decrepitude. No, the mice live a longer but much more healthy and agile life, with a reduced incidence of certain cancers and other diseases.

One test of biological aging placed two mice of the same chronological age on top of a rolling dowel. The old, graying mouse that had eaten as much as it liked lost its footing and plopped off the dowel long before the calorie-restricted mouse, which continued to dance away on top of the rolling log.

Roy was particularly excited because no one had ever successfully studied a human population on a rigorous hi/low diet (high nutrients, low calories). It is not an easy diet to follow, and subjects cheat. But there was no way we could cheat, so Roy hoped to find out if humans showed the same biological response to the diet as mice. Expected effects included a reduced fasting blood-sugar level and reduced blood pressure.

Every eight weeks since Closure, each of us had fallen victim to Roy's medical probings. He poked and prodded us, took our temperature and sampled our blood, weighed us, pinched our skin with freezing calipers to measure its thickness for body fat calculations, took pictures of us naked to compare morphology before weight loss and after.

It appeared that we were reacting as his mice did, just like other rodents and monkeys that had been studied. So, while our stomachs complained, we could console ourselves with the thought that we were probably getting younger and healthier with every passing day.

We were, however, becoming ever more obsessed with food. After dinner some of us sat around the cold slab of granite that was our dining table and engaged in a recurring form of therapy—food fantasies. We imagined and described in exquisite detail a rapturous meal we wished we were eating. Sometimes I could smell a flourless chocolate torte as I brought the empty fork to my mouth. As I placed it on my tongue I could feel the gooey, creamy consistency, and taste the full, rich, pungent, dark chocolate, and I washed it down with an imaginary cappuccino. Or I would sink my teeth into an exceptionally ripe, sweet, fantasy strawberry, the aroma wafting up my nostrils as it came close to my mouth. It sounds masochistic, but we imagined the food so vividly it was almost as if we had eaten it. Our hunger was not only for quantity, but for the delight of a good meal with a luscious glass of wine. We were hungry for stimuli.

When watching a film, I would focus on the eating scenes, forgetting the plot. Taber and I would talk about what was on movie plates and in movie glasses. We took to watching cooking shows on TV. Once we so wanted to taste the shrimp and melon balls that a Hong Kong chef was cooking that we phoned and bought every book he had written, although we could not possibly use the cookbooks for another eighteen months.

Margret had placed a hot-dog stand not far from the Biosphere to serve the thousands of hungry visitors. Sometimes we lined up in the second story windows of the Habitat and took turns peering through binoculars at fat people (for everyone seemed overweight to us then, even the slender people) who were spurting ketchup on sausages and shoveling them into their mouths. We were culinary voyeurs.

But despite my hunger, I could not embrace the tradition of eating insects. Mark claimed to have tasted the crickets that were chirping throughout the Biosphere. That sounded quite disgusting and, to my English sensibilities, symbolized stooping far too low, tantamount to admitting we were scrabbling for our survival.

Even so, we had numerous discussions about whether it would be remotely possible to turn the enormous cockroach population into food.

We had intentionally introduced five species of roaches into the Biosphere to recycle leaf litter in the Wilderness biomes. None of those was a problem. But a sixth, uninvited stowaway species had snuck in, most likely on a plant—an Australian roach. All the organic matter throughout the Biosphere seemed a perfect habitat for these bugs from Down Under. There were literally millions, many in the kitchen. It did not help to know that roaches are an unsolved problem in many large botanic gardens—at night, the floor of the Palm House at Kew Gardens near London is thick with them.

Although we kept everything spotlessly clean, once the kitchen was plunged into darkness, hordes of insects turned the white countertops brown. Eating them directly was out of the question. So the person on night watch had the chore of creeping into the kitchen to catch them unawares. Armed with a vacuum cleaner, he or she flipped on the light and vacuumed up as many roaches as possible before they all scuttled away. We then carried the roaches down to the animal bay and fed them to the chickens, who, although startled awake, leaped into action chasing after the bugs, which were a great source of protein. Thus roaches were transformed into a sparse supply of chicken eggs.

We also trapped roaches. Roy had ordered KY Jelly for gynecological exams, but instead of four tubes, he received four cases. So we smeared it around the rims of containers. The roaches scuttled in to eat the piece of papaya skin awaiting them in the bottom but could not escape over the slippery jelly. They, too, became chicken feed. I gained a certain amount of perverse entertainment from the notion that the biospherian "heroes" had to fight insects at every turn for our place in this new world's order.

Out of this escapade "evolved" a whole new species of insect, never before seen on planet Earth. We had often discussed the possibility of new species evolving inside Biosphere 2 after one hundred years of hermetic closure. The Biosphere was potentially a deathtrap for some species confined to the small world, or a crucible for new species, giving scientists the chance to capture evolution in progress. Dr. Stephen O'Brien, member of the Science Advisory Committee and acclaimed geneticist with the National

Institutes of Health, had initiated a project that would track the genetic diversity of particular species inside. Eventually, cockroaches were chosen as the initial species to follow, and some of those trapped by the KY Jelly were dedicated to scientific research.

One evening Roy preempted the evolutionary geneticists, announcing that he had discovered a new species—the Cockroach Butterfly. With the wings of desiccated lablab bean leaves and the bodies of dead roaches, Roy Walford transformed two ecologically useful but disdained organisms into a mythical creature. He carefully pinned his holotype (the original specimen against which all future specimens would be identified) as an entomologist would, the animal's abdomen stuck to the top of a long steel needle and the pin skewered into a piece of foam. On the clinic's bookshelf sat our doctor's new discovery, his contribution to the world of art.

I loved that Roy used his science background to poke fun at our assumptions and at our idealistic notions of a utopia in which we would leave all the scourges of the planet outside the heavy steel airlock.

Intrigued by the idea that there were now two distinct biospheres communicating with each other about matters of science and management, in early April I organized an Inter-biospheric Art Festival to expand that communication to include poetry, music, painting, and film. All eight of us crammed into our videoconference room and exchanged our creations with the audience and performers in Biosphere 1, who in turn read their poetry and played their music. It was an evening of warm exchanges with a sense of community, even between the eight of us and Mission Control. Every member of our crew read, played, or showed some creative piece, and no matter how guarded it may have been, a little of each person's soul was laid bare that evening.

A few evenings later I was at my desk in my room when I heard a knock on my door and Taber delivered the bad news.

"It looks like we may be losing oxygen."

Chapter 14

GOLD OR LEAD

B y now, roughly seven tons of oxygen was missing. And we had no idea where it had gone.

Maybe we could find the problem and fix it. But maybe it was unfixable, even if we could find it. If we were lucky, the rate of loss would slow down, and maybe we could adapt like mountaineers somewhat do, and squeak through the two years. But if the oxygen was disappearing at a constant rate, at some point we would either have to leave the Biosphere, ending the experiment early, or bring in oxygen, breaking the promise that no material would go in or out of the enclosure during the two years.

It was hard to imagine that only a couple of weeks earlier we had celebrated six months of closure with a PictureTel link to the famed Explorer's Club Annual Dinner. The Club had received us into its ranks with the likes of Sir Edmund Hilary, who first climbed Mount Everest, and Sylvia Earle, who explored the deepest regions of the oceans in tiny submarines. With pomp and ceremony the club had presented us its flag to fly outside Biosphere 2 until the end of the mission. It was a great honor.

Now, I wondered, were we about to let them all down? Could the mission actually last the two years? To abort it would be a terrible humiliation.

I looked over at Taber, and noticed that his eyes, though lined with exhaustion, were shining with excitement. "This is the gold of this project," he said. I couldn't imagine what he was talking about. Gold? This was lead. How could anyone make gold out of it?

"Now," he went on, "we have a problem to really sink our teeth into—something we didn't predict. We can kick the scientific machinery into high gear and see if we can solve this. We get to test the hypothesis that the Biosphere will not fail for some reason that we cannot explain or find a solution for. It tests the assumption that we would be able to understand the Biosphere and its complexities. This is the reason we're in here. Now we can show that we're about real science."

What he said was true, but we were still in shock. We all knew there were tons of oxygen missing. Lost. Gone. Unaccounted for. Taber would tell me later how he walked through the Biosphere looking everywhere, as if he would suddenly come across a huge pile of oxygen. He wondered where more than seven tons of oxygen could possibly be hiding. In the soil? In our Ocean? He simply could not imagine a place that it would be sitting all alone. It had to have reacted with something, but what? Carbon, making carbon dioxide, was the most obvious answer, but where on earth could ten tons of extra carbon dioxide be locked away?

Over the next few weeks Taber analyzed the atmosphere about every three days and pieced together data from old samples. He discovered that the oxygen was slowly, steadily decreasing at 0.23 percent per week. The news could only have been worse if he had found it was decreasing faster and faster.

Meanwhile the news had shaken Mission Control. John and Margret imposed a gag order, demanding that the entire problem be kept under wraps until we knew what had happened and why. Taber was soon involved in a scheduled videoconference with a group of scientists who listened to presentations about the experiment thus far. One of them had made some studies of the oxygen cycle in tiny sealed environments—essentially, mini-biospheres and he inevitably asked Taber, "How's the O_2 level?" Taber could not answer.

I could not imagine what this news was doing to John, who always appeared completely convinced that we would breeze through the two years without any major unforeseen problems. He had talked as if nothing would seriously jeopardize the validity of the Biosphere.

He apparently was denying the mystery oxygen loss completely. Finally, a couple of weeks later, he delivered a report to the Science Advisory Committee that claimed the oxygen would level off at 16.25 percent. Some of us were flabbergasted. Freddy had drawn the most optimistic possible curve through the handful of data points on the graph, pushing the data to an extreme to show a leveling off, and thus complete atmospheric recycling after an initial settling period. But a line that showed an almost linear decline with no end in sight was far more believable.

The report also intimated a deliberate "oxygen program," which was reverse engineering at best. Oxygen was an afterthought, as it had never been a problem in any previous experiments in the Test Module. That is why there was so little credible data thus far in the mission. All in all, the report was a heap of ass-covering with a good deal of denial thrown in. It certainly was not open, honest scientific discourse.

Shortly after discovering the oxygen loss, Taber had painstakingly written a document showing that the slow decline of oxygen was likely due to the production of carbon dioxide in the soils. As microbes broke down the compost in the deep soils, they took oxygen out of the atmosphere and produced carbon dioxide through respiration. It boiled down to a simple error in design—we had put too much compost in the soils. A nagging question still remained, however: Where was all that carbon dioxide?

Taber had calculated how much carbon dioxide was stored in the scrubber, in the heaps of biomass in the basement and elsewhere, to show exactly how much carbon dioxide, and therefore oxygen, was missing. He even had an initial hypothesis of where it was tied up; he suspected that the carbon dioxide was reacting with calcium in the Desert soils to form carbonates.

Roy, Linda, Taber, and I were all furious that John was not telling the whole story to the Science Advisory Committee. The other four biospherians seemed to back up what I frankly considered management's dishonest behavior. It was unbelievable to me that they—members of the crew who had to suffer the consequences of oxygen loss directly—could excuse management's mendacity. Did they really think they could fool the SAC?

Gaie, as head of research in the Biosphere, and Mark, who served on SBV's board of directors, had ample power to do something, but they didn't. Taber talked with Gaie immediately upon hearing about the ridiculous claims John's report made, telling her they were entirely false. She shrugged.

Meanwhile, Mission Control issued an edict that Linda, Taber, and I were not allowed to talk to the Science Advisory Committee. There was nothing we could do but stew among ourselves.

We were soon placated (at least somewhat) when the Science Advisory Committee and John invited Dr. Wally Broecker of Columbia University's Earth Studies Institute to help Taber figure out where all the oxygen was going—and he accepted the task.

Wally was famous in the world of global research, for he had first postulated the potent role in the world's climate of the Atlantic Ocean's great thermal conveyor, the Gulf Stream. He was just the sort of global thinker to help unravel the enigma of our oxygen cycle.

Our situation was being taken seriously, and would be dealt with in an appropriately scientific manner. But as bad handling of the media and now the absurd misinformation to the Science Advisory Committee unfurled, the crew split apart with striking symmetry. Two men and two women lined up on either side of the burning issue: what to do about the missing oxygen. At some point well before the scheduled end of our mission, there simply wouldn't be enough to support the crew.

We had suffered the same division when we realized that we were not going to have adequate food. Roy, our physician and the most credentialed scientist on the crew, suggested we compromise some of the material closure of Biosphere 2 by bringing in a measured amount of extra food, such as cooking oil, so that we would have more energy. We would thereby be able to devote more time and effort to solving the pressing problems of our life support system, and address one of the greatest critiques of the project by performing more science.

After all, this was in effect the shakedown cruise of the Biosphere, and it seemed a terrible waste of time and money to simply continue without understanding and fixing the problems at hand.

At first I had been in the management camp, which maintained that if we got too hungry, we should send people out rather than bring food in. We had stated that we would not move anything in or out during the two years, and darn it, that is what we would do. We did not need to bring in more food just because we were hungry—that was a show of weakness, and Lord knows, we had been catching it from the press recently. How would we explain this one? I could see the headlines: "Biosphere Fails: Mounds of Food Shoveled in Airlock."

From a purely scientific point of view, this was, of course, ridiculous. It made absolutely no difference to the outcome of the first two-year experiment whether we had grown 80 percent of our food, or 90 percent. We simply needed to know exactly how much we brought in, how much we ate, and the contents of the food.

Thus, I was eventually persuaded to take the other side of the argument—the scientific side—concluding that the others had fallen prey to the tyranny of the Biospheric Ideal of Perfect Material Closure. I became convinced that their logic was based on symbolism, not science. They had completely forgotten that (a) this had never been done before and thus was highly unlikely to be perfect the first time, and (b) we were running a scientific experiment, not an exercise in survival.

With the oxygen crisis, the two camps became entrenched. The science division wanted to betray the goal of complete material closure. They, the other four crew members, and John and Margret (a.k.a. management), preferred to sacrifice scientific rigor to what clearly had become an impractical goal.

And they were willing to do that because of what? They were afraid of what the press might say? It went against an impossible definition of success? It seemed so arbitrary.

What stung most was that I found myself pitted against the very people to whom I had pledged allegiance only a few months before, people who had claimed to have the best interest of the project in mind. I was miserable. At that moment I knew Sartre was right— Hell is other people!

Actually, I believe Sartre is incorrect. There are those unfortunates who have an objectively terrible life, but generally we create our own hell. One of my favorite sayings is by Oscar Wilde, "My life may be miserable, but I am not." It is all a matter of perception and how we choose to respond to a situation. Inside Biosphere 2 we were making ourselves thoroughly wretched.

Chapter 15

LINES DRAWN
IN DESERT SAND

S urprise! Rather than delivering the final blow to Biosphere 2's credibility, the riddle of the missing oxygen brought the scientific community flocking to us. Here was a fascinating issue for everyone to think about and talk about. They could discuss its ramifications. They could postulate hypotheses. Some scientists were flabbergasted that we could even measure such a slow decline, and finally believed in the structure's tight seal.

At the end of April, we held a closed-system workshop. We presented the results and issues of Biosphere 2 research to date to thirty scientists from the United States, Japan, Europe, and Russia. We focused on our carbon cycle (and the related oxygen cycle) and its relevance to sealed environments in general as well as to global warming on Planet Earth.

The interest at this workshop and others, which included marine science, Wilderness biome ecology, and medicine, confirmed that Biosphere 2 was a serious enterprise and that the crew inside was not a bunch of amateurs. The project was moving in the right direction scientifically.

The Science Advisory Committee was expecting to receive a full research plan by early July. It has always been an enigma to me why we did not have a research plan prior to starting the two years. I suppose it was because we were all so focused on finishing

Biosphere 2 that there was no time to concentrate on what happened after September 26, 1991.

Writing the research plan gave an opportunity to show off Biosphere 2's scientific and technical prowess and put a lot of jaded criticism to rest.

Instead, the research plan ignited a huge controversy.

In 1989, Ed had endowed the Chair of Biospherics at Yale and to this day, the university runs the Yale Institute for Biospherics Studies. Concerned by the increasing separation between fields of science, John and Ed had wanted to bring diverse disciplines together under the umbrella of a single, all-encompassing science that studied materially closed systems and their cycles: *biospherics*. The new field of science would encompass everything from miniature systems and Biosphere 2, to the Earth's biosphere.

But John and Ed were trying to counter more than forty years of division and antagonism between big-picture and small-picture scientists. In the 1950s, biology split into microbiology—the favored child—and what the microbiologists considered the dubious branch, evolutionary biology.

In his 1994 book *Naturalist,* Edward O. Wilson explained that the fundamental difference between the two was that the microbiologists took a "bottoms up" approach to studying life. They studied life in the smallest possible unit, such as RNA and DNA, and worked up to studying cells, then organisms, followed by interactions of organisms in ecosystems. The evolutionary biologists took a "top down" approach, postulating that all life could not yet be understood solely through biochemistry, and that studying the birds, the bees, and the principles distinctive to organisms and ecosystems was paramount. This division survives today.

Biosphere 2 was the big-picture ecologists' dream project. It would help elucidate a set of "biospheric laws" that could be applied to Biosphere 1. In a 1979 paper entitled *A Foundation for Ecological Theory* published in the book, *Biological and Mathematical Aspects in Population Dynamics,* Dan Botkin, a leading ecologist and environmental scientist now at the University of California at Santa Barbara, and associates wrote, "Suppose ... we find a set of control policies for a set of goals for one ecosystem and then are confronted with a second ecosystem and a second set of goals. If we can find a way of generalizing from the first to the second situation, then we have formulated a kind of ecological law."

The idea was captivating. In a commentary to *Science* in March 1993, Eugene Odum wrote, "The experiment [Biosphere 2] is not traditional, reductionist, discipline-oriented science, but a new, more holistic level of science that might be called biospherics, the name selected for a new research center at Harvard that was inspired by the current experiment in Arizona." Biosphere 2–along with Eugene and Howard Odum, Stewart Brand, and others–was attempting to transform a fairly widespread if guerilla-like movement into reliable, scientific action.

Nonetheless, when we were designing and building Biosphere 2 in the 1980s, ecology was pooh-poohed as a "soft science" and a tool of alarmist environmentalists by many scientists, rather than seen as a new, burgeoning science. Evolutionary biology, of which ecology was a part, was still considered a second-rate branch of biology, destined for musty museums, cramped backrooms, and small budgets.

Dr. Howard T. Odum, Eugene's brother and one of the fathers of systems ecology, wrote in a 1996 paper in the journal *Ecological Engineering,* "The self-organizational process of Biosphere 2 was a beautiful living model with which to study aspects of the large earth by comparison, but when journalists asked establishment scientists, most of whom were small-scale (chemists, biologists, population ecologists), they got back the small-scale dogma that system-scale experiments are not science."

From a reductionist science point of view, Biosphere 2 was too complex to perform rigorous science, and there was only a single system to study, with no control and too many variables. From a holistic stance, we could study Biosphere 2 in much the same way as climatologists and ecologists can study the overall operating system of the earth, but with one, very important difference. Unlike the earthly biosphere, we knew exactly what had been put into Biosphere 2, and since it was hermetically sealed, nothing could have entered or exited without being recorded. Thus, in theory, we should be able to track every atom through the Biosphere's systems, providing us with a detailed map of their cycles.

As Taber recognized, the oxygen loss would certainly test whether this was a valid assumption, or whether the cynics were right in their insistence that Biosphere 2 was just too complicated, with too many variables, to do reliable science.

In fact, we required the two approaches of science—both the bottom-up, reductionist approach, along with the top-down, holistic view. As Dr. J. Hart wrote in 2002 in *Physics Today,* "Physicists seek simplicity in universal laws. Ecologists revel in complex interdependencies. A sustainable future for our planet will probably require a look at life from both sides."

Without reductionist science, Biosphere 2 would not have been built; reductionist and holistic science working hand in hand would solve the oxygen-loss riddle.

However, instead of touting biospherics as a vehicle to bridge the gap between both forms of science, John turned it into a fight for holistic-versus-reductionist science, which he held in considerable contempt, labeling it backward, linear, and narrow-minded.

The Biosphere 2 research plan brought this conflict front and center. It was the first attempt at putting together such a plan under the aegis of what could be called biospherics. Gaie, as the head of research inside Biosphere 2, was charged with putting together the plan's first draft, with help from Roy, the most credentialed scientist on the team. He knew what a research plan should look like, and he knew the ins and outs of the scientific community and its politics. If anyone at the project knew how to sell the idea of a new science of biospherics it was Roy, along with the Science Advisory Committee.

Most of us worked feverishly to finish writing up research projects for a "taster" that was to go to the Science Advisory Committee for feedback. We had already embarked on more than fifty research projects, which included collaborations with top scientists from the Smithsonian Institution, Lamont Doherty, George Mason University, University of Arizona, Yale School of Forestry, Georgetown University, College of Charleston, Monnel Chemical Research Institute in Philadelphia, University of Texas, Michigan State University, University of Georgia, and Cornell University.

As would be expected, much of the initial research focused on understanding the workings of the new world we had created, though some had direct applications to Biosphere 1, particularly in the area of restoration ecology.

Our research discovered a creature hitherto not seen in nature, *Euhyperamoeba biospherica,* named for Biosphere 2. Don Spoon, a member of the Ocean's research team, first discovered the marine microbe. It was later found to exist in almost undetectable quantities in Biosphere 1 oceans. The environment in the Biosphere's Ocean was apparently perfect for the amoebae's reproduction and survival, because no predators were eating it.

Researchers and their projects live and die in the scientific community by the number of papers published. Despite all this work—and even though the oxygen loss had garnered honest interest—we were still far from being a model of good science. Some information was not published out of concerns for confidentiality and protecting corporate assets in the form of information. Nonetheless, no matter how valuable a project Biosphere 2 might be, no matter how hard we were working inside, no matter how incredible a feat it had been to build it in the first place, if the scientific community damned the project, Biosphere 2 would have one hell of a time recovering.

The life-sciences editor for Springer-Verlag publishing company had just refused publication of an appendix about Biosphere 2 in Howard Odum's new book on mesocosms because, as he stated in a letter to Professor Odum, "Based on my own observations and on discussions with other scientists, I have a lot of concerns about whether the Arizona group is conducting legitimate research or is more concerned with self-promotion and finding ways to slip pseudoscience into the literature." He did eventually agree to include the appendix in the book. Still, it was a damning statement and demonstrated how tenuous the project's position was in the scientific community.

But according to my journal, John at this time was threatening to send only a list of projects to the committee instead of a detailed plan, despite the fact that several detailed experiment plans had been written up.

It made no sense to antagonize the Science Advisory Committee. John could not win. They were all top-flight scientists and they had their own reputations at stake. They provided the project's credibility at that point because there was little else to pin it on.

I was livid with Gaie for not making John see this. Now, I am sure there is little she could have done, but then I was beside myself that she, Laser, Mark, and Sierra were all lining up behind John. I kept asking myself why they could not see that the course they were taking would likely result in project murder and career suicide.

I wrote page after page of diatribes in my journal during those days, but in truth, my condemnation of my fellow inmates was less about them, and more an exploration of my own thoughts and feelings about John. It seemed to me that many of his actions during the past year had belied the self-examination, behavior modification, and "working on oneself" he taught. The times— in the plural—before Closure he had run out of the room yelling, "I'm resigning," hardly seemed the act of a mature leader. And now his refusal to cooperate with the Science Advisory Committee on the research plan seemed asinine.

But for ten years, I had believed him infallible, held him in such high regard that nobody could have lived up to the godlike image. When, as I see now was inevitable, he showed himself to be chipped, cracked, less than perfect, humanly flawed, my adoration turned to hate.

I am sure he was just as disappointed in me. I had not stuck by him as I had promised. But how could I stand by him when he was betraying not only me, but hundreds of people who had sunk their passion into the project for years? I was struggling to come to grips with the fact that I was losing a dear friend, which I had considered him for more than a decade before the airlock closed on Biosphere 2.

I had loved so much about our way of life together in the good ol' days and I was determined to sort through my thoughts and feelings of ten years in the Synergias and not throw out the good with the bad. At times, I was so furious I thought my head would burst with anger. At others, I felt like my heart was breaking.

Before we even entered the Biosphere, John had asserted that there would likely be a schism in the crew, but he had foreseen things going quite differently. He had surmised that Roy and Linda might not line up behind him and tasked Gaie, Laser, Taber, and me with keeping the crew together. We were supposed to have been the glue. The four of us were close friends and this scenario seemed reasonable.

But, at the end of February 1992, Gaie and Laser approached Taber and me about putting some money into a hotel in Belize. We were to purchase one of the cabanas. Taber and I decided that it was not a good investment of the little money we had, and said no. Instead, we decided to buy a piece of land that we could see with binoculars from the library, and to build a home with generous help from my parents.

With this action we declared our independence from the group. We were no longer going to live within the compound, part of the traditional setting of the Synergias.

Prior to this decision, I had spoken with John several times a week. After telling him of our plans in March, I spoke to him only a couple of times. Gaie and Laser never again came over to our apartments for a friendly chat over a glass of red wine from one of the twelve bottles I had finally mustered up the courage to smuggle in before the metal door clanged shut on us all.

It was if we were starting the process of divorce after a passionate love affair had gone awry, each party feeling betrayed by the other. As scientific credibility became a growing issue, the two camps solidified as Science versus Management. In a paper Roy later published on biospheric medicine, he characterized the split as two factions with "one totally loyal to and supportive of outside management decisions, the other . . . increasingly hostile to what they considered an arrogant, scientifically inept, and abusive management team."

But the crew still did not acknowledge our social problems openly. I think that each crewmember honestly wanted the eight of us to get along, to be a strong team, bonded in adversity. At the outset, I naïvely thought that, by necessity, life inside Biosphere 2 would pull us together in a sort of utopian group, living in harmony. Instead, it seemed that the adversity inside Biosphere 2 was dynamiting our previous friendships and faith in one another.

We just could not admit that the Synergia lifestyle had reached its limit. We had realized the dream of building a new world that we would infuse with our own Synergist values. It would be the fruiting body that would eject seeds of our ideals throughout the world. Here we were, ambassadors of our species, licking plates, throwing tantrums, rebelling, and emotionally torturing each other. Our "civilization" was wracked by power struggles and petty infighting, like so many other groups.

We all put on a brave face most of the time, and at parties we tried to act as if nothing was coming between us. Earlier in the year we had held a campfire picnic on the beach for Laser's birthday, where we all sat warming ourselves by the light of a fireplace on video. Mine had been in our small gym that nobody used. Everyone came dressed as bodybuilders, despite the fact that our muscles were hardly giant, throbbing masses beneath our thin skin.

Sierra's birthday on June 15 was a jolly black-tie dinner. Everyone dressed up. But all the guys' jackets and dress shirts hung off them and made their necks look even scrawnier than usual, swimming around inside giant shirt collars. Summer solstice came and we all lay around on the carpeted floor in a Roman feast, leaning luxuriantly on cushions and eating until our waistbands were too tight and our bellies ached.

For Roy's birthday on June 29, we held a *lungi* (a bright cloth wrap usually worn by men around their waists in parts of Asia) party on the balcony overlooking the Agriculture. All the men sat around bare-chested, while the women found creative methods of covering their chests with cloth and long wigs.

One evening we watched a video recording of the play we had all written and performed before Closure, *The Wrong Stuff.* We all laughingly commented on how prescient it had been, predicting food shortages, disgruntled biospherians, and now even a possible mutiny against the management. Seeing ourselves acting on the screen gave us enough distance to laugh, and a sense of security, knowing that the action we were watching was on a stage.

We had not had acting class for months, and I think the primary reason (although the justification was that everyone was too tired) was that we did not want to explore our deep divide and unleash the issues, even on stage, as they were more explosive than any of us felt we could handle in the open. We had to keep them bottled up. Watching the video of *The Wrong Stuff* was the closest we had come to admitting that we had serious problems with our group dynamics.

And where, I wondered, were our leaders, the empowered group of galvanizers they had been prior to Closure? Neither one ever brought the team together for a roof-raising, morale-boosting pep talk. In personal correspondence in 2000, John explained that he and Margret had spent at least a hundred hours in 1991 deciding whether to tell the staff what was going on—that there was a power struggle with Ed. But every time they decided not to, as they were convinced that Ed would "start to talk to us again and not to make the split inevitable." So Ed was the demon here? And where was Ed, anyway? Why wasn't he rallying the troops to soldier on? Ed was busy with projects in Fort Worth and around the world. In fact, I felt abandoned because in two years, Ed never spoke to me, except at the first-year anniversary celebrations.

He did, however, notice that the project was hemorrhaging cash. Although the four of Us on the inside had no idea of the financial problems, Ed began pressuring John and Margret.

For some reason I will never understand, John felt betrayed because Ed wanted him to get the project's financial house in order. He thinks that as early as 1989, Ed began a concerted campaign to drive John and Margaret out of Biosphere 2, so Ed could take over their share for free. In a letter dated August 2, 2000, John told Rebecca Reider, who was writing a thesis on the project, that Ed had gotten scientists, media, and various biospherians—including me, Linda, and Taber—to join his takeover campaign. "When Ed developed a separate policy to take over Biosphere 2 by forcing us to sell, this meant that Margret and I had to take the rap for changes he promulgated, in exchange for him authorizing ongoing projects we considered essential. . . . Roy was not far off when he said I looked a bit crazy, making change after change in 1991. . . . Margret and I took the blame for suddenly being on people's backs more than before in having to accommodate our partner's changed world view." They went along, John said, in order to keep the money flowing.

In fact, I thought this whole mess was as much Ed's fault as John's. "Why doesn't he just get rid of John and Margret?" the four of us would sometimes ask of each other, baffled at Ed's seeming passivity toward what I perceived to be their existential ineptitude. Now I believe they were still there because they had a business agreement, and almost fifteen years of complicated business and personal relationships.

So I sided with the SAC, which John considered tantamount to siding with Ed. Back then, I knew there was more going on than I could see and hear, but I made my decision. The project had to be credible scientifically, otherwise it would cease to be important, and John and Margret were not showing themselves up to the subtleties and complexities of that fight. I am still convinced I made the right choice.

It was a crime that I had to make a choice at all.

I did not allow myself to think about the specifics of what would happen if we popped the cork on our emotions. I simply did not know how to handle so much confusion, so much hurt. Taber and I sat up late discussing the awful situation, and more and more frequently we talked things over with Linda and Roy. We always concluded that it was best for Us to "keep the lid on." But so much buried emotion festers, and it festered in Biosphere 2.

The sun blazed across the Biosphere the latter half of June 1992. Our oxygen had remained level for a few days and the carbon dioxide was the lowest it had been in eight months—500 ppm. Everyone continued to work hard harvesting rice, planting sweet potatoes, weeding, giving interviews, weeding, fixing the wave machine, weeding, writing up experiment plans, weeding, and butchering piglets. And weeding.

Linda and Mark were studying the functioning of the Wilderness biomes, such as how quickly leaf litter decomposed in every Wilderness biome but the Ocean. The decomposition of leaf litter by insects, bacteria, and fungi is an important part of ecosystem nutrient cycling. The journal *Ecological Engineering* published a paper by Mark in 1999 indicating that leaf-litter decomposition contributed approximately 7 percent of the atmospheric carbon dioxide.

Dr. Phil Dustin from the College of Charleston, South Carolina, and an authority on coral reef ecology, worked with Gaie on comparisons between a Mexican reef and the Biosphere 2 coral reef. Mark, Sierra, and I were writing a technical paper on the design and operation of the Intensive Agriculture biome for publication in *Outlook on Agriculture,* an international journal. Linda and Roy had started studies of the changes in genetic variation and genome structure and organization, using the cockroach as the sample organism.

Roy and Taber continued work on their ongoing blood toxicology study. Every two months, each biospherian rolled up his or her sleeve and Taber took a sample of our blood and froze it. Roy sent it out for analysis at ACCU-CHEM Laboratories in Texas, which Taber had found to sponsor the research. The hypothesis was that the fat soluble toxins—that is, those that would be released from our fat into our blood as we lost weight—would increase, remain relatively stable while our weight remained constant, and then reduce again as they were reincorporated into our fat as we gained weight after Closure. It was thought that many of these compounds are not metabolized and excreted by our bodies very well, and this would be verified if the levels remained unchanged while our weight was constant. This research could only be done in Biosphere 2 as there were no outside sources of the toxins to muddy the data.

At this point, the results already were ghastly—most of us had significant concentrations of DDE, DDT, and other compounds that had not been used for years. Roy and Taber, along with Dennis Mock of UCLA School of Medicine, and John Laseter of ACCU-CHEM, eventually published the results in the journal *Toxicological Sciences* in 1999, showing that lipophilic compounds (those stored in body fat) DDE and PCBs increased in our blood as we lost weight and decreased as we put the weight back on after the two years. The paper concludes that, "[A]lternating periods of weight loss and weight gain (so-called yo-yo dieting), with flushing of lipophilic toxicants in and out of the body, may be harmful in terms of internal exposure to toxic substances."

Roy was forever inventing pranks for us. One afternoon, he, Taber, and I marched into the Rainforest mountain, tore off all our clothes and painted each other's snow white skin with bright body paints. Looking like New Age tribesmen, we wandered naked through the trees and vines of the Rainforest, bare feet caressing the leaf-covered ground, careful to stay out of public view. We were the natives of our modern high-tech world, exploring our direct relationship with nature while enjoying the benefits of modern technology.

John Denver made a surprise visit. Members of the European Space Agency came by. The producer for Pink Floyd graciously listened to Taber's and my music. One Sunday afternoon, Bill Mathews sat on a chair outside the visitors' window under the hot sun, his fingers flawlessly playing passionate classical Spanish flamenco, while Safari held the telephone to his guitar.

On July 4, all eight of us climbed eight flights of spiral staircase to the circular library, the highest point in the Biosphere, where we watched fireworks while relishing doughnuts that Sierra had made. For a couple of hours it was as if we were all best friends. We ate, chatted, and laughed.

But fireworks were going off in and around the Biosphere over rumors that the SAC intended to replace John with a new scientific director. Roy communicated frequently with Tom Lovejoy, and he assured us the rumors were true. The four of Us—Roy, Linda, Taber, and I—were relieved to hear the news, but we imagined that the four of Them—Gaie, Laser, Mark, and Sierra—must have been livid. By now, no honest conversations were possible between Us and Them.

There is a line beyond which there is no return and friendships are irreparably damaged. On July 5 that line was crossed, and I knew that I no longer needed to feel guilt over my failing friendships, nor could I try to repair them.

Some semblance of a research plan had finally been assembled and sent to the members of the Science Advisory Committee. In turn, Tom Lovejoy sent it to Roy along with the accompanying cover letter, which Roy thought he had helped Gaie write. What Roy showed us did not resemble in any way what he had written. Roy, Taber, and Linda had written many of the research projects and had expected to be principal investigators of Biosphere 2. Not one of them was listed as a principal investigator. The only names mentioned were John, Gaie, Freddy Dempster (the chief engineer), and Norberto Alvarez Romo (the head of the computer systems, and Margret's boyfriend). Not one of them had a Ph.D.

I have rarely seen Roy angry, but on that occasion he was spitting, his lips pulled back over his square teeth, repeatedly snarling, "Those motherfuckers!"

Roy had put his own expertise and credibility on the line in the Research Plan. Without Roy's name on it, the entire plan lacked credibility—he was the only crewmember who held a doctoral degree at that time. Linda was the next most qualified. She held a B.S. in botany and field ecology and an M.S. in range management, had a wealth of field and research experience and was well respected in the botany and ecology community in Arizona where she had worked for a number of years before arriving at Biosphere 2. She had been: a researcher at Miami University in Ohio studying methods for prairie restoration under a grant funded by the National Science Foundation; the field coordinator and botanist for a Washington state-funded investigation of the feasibility of reintroducing gray wolves to the Olympic Mountains of Washington State; a field ecologist for the Evergreen State College, Olympia,

Washington, on a project to formulate resource-management alternatives and recommendations for the Alaska Peninsula.

Gaie also had a background in science, with a B.S. in biology and an M.S. in environmental studies from Yale, and a great deal of field experience. She had been at sea on and off for over ten years, participating in research for the World Wildlife Fund and the United Nations Environment Programme on cetaceans. The rest of us either had inapplicable degrees (Mark held a B.A. in philosophy from Dartmouth College, for example), or no degree at all.

But Biosphere 2 was privately funded, not a government-sponsored project, which afforded us a certain amount of freedom. The project's management could put whoever they felt appropriate inside Biosphere 2. Few of us inside Biosphere 2 claimed to be scientists—we were managers. In essence, we were capable technicians. But, people generally see what they expect, and people expected that the crew of Biosphere 2 would be scientists or engineers.

The SAC could have advised putting more professional scientists on the biospherian team but did not, although they did recommend including more "practicing scientists with advanced degrees" on staff outside. This was principally, I feel, because they knew that Biosphere 2 needed more than scientists. The two-year experiment required people well versed in Biosphere 2's operations and its mechanical systems and well-trained technicians, which all of us were. Also, they knew that there were few people willing to incarcerate themselves for two entire years. As our closure wore on, the steadfastness we showed in the face of so much adversity led many skeptics to respect us personally, if not our science.

Ten months after Closure, the four of Us would huddle at one end of the dining table for lunch, and the four of Them would either leave and eat elsewhere, or crowd together at the other end of the table. Sit-down dinners gave way to people grabbing their food and running to their rooms. We still held Sunday-night speeches, but the four of Us were showing up less often. The four of Them began holding Thursday night philosophy evenings together in one of their rooms.

One afternoon, Taber and I were walking along the hallway to the dining room. Gaie and Laser were walking toward us. As we passed them we hugged the wall, and averted our eyes. So did they. That was the way it was for the remaining fourteen months. We never looked each other in the face again.

With nearly palpable hostility hanging in the Biosphere like a toxic cloud, Taber and I looked for ways to escape. We watched totally unmemorable TV programs. Sometimes we hiked the two hundred feet down the stairs and through the basement to the beach and sat looking out through the spaceframe at the stars. We would search for Mars in the night sky and contemplate how different things might be if our Biosphere were suddenly transported there and we were looking not at the tiny red speck of Mars, but at the bluish dot of our home planet. Perhaps on Mars, with the safety of home at least forty-eight million miles away, we would have been able to pull together.

But then again, perhaps not. I wondered if people are really meant to be enclosed in small spaces, even as large, beautiful, and varied as Biosphere 2. The human species, after all, did not evolve indoors.

The enclosed life was often one of a monastic sort of contemplation. Not only were we food-limited and increasingly oxygen-limited, we were impression-limited. The simplicity of life was at once fiercely liberating and crushingly oppressive.

In spite of the controversies and turmoil, my brain was not bombarded with the millions of stimuli we all experience in modern life. I could not distract my mind by driving into town to see a movie, go to the theater, have a fancy dinner, or go dancing.

Despite the melodrama, the "social spaghetti," as Roy called it, was vastly simplified. I dealt face to face with only seven other people, and my interactions with those on the outside were largely filtered through Mission Control. I did not have to deal with many of the day-to-day complexities of the world outside. Nope. I had to find my entertainment, my satisfaction from within the confines of the 3.15 acres of Biosphere 2 and from within the thirteen hundred cubic centimeters of my brain.

I found this the most difficult thing about our enclosure, being someone used to looking outside for satisfaction. From the age of eighteen, I had been riding a wave of constant external stimulation until the day the door on Biosphere 2 slammed shut. I had had almost no time or energy for introspection.

I had almost never lived alone, been alone, to face myself and my own angels and demons.

Now I was thrust into the position of staring inwards. Looking straight into my own soul was at best uncomfortable, at times excruciating. I forced myself to be honest with myself, although sometimes I did not like what I saw. Sometimes I felt only emptiness. Other times, toxic thoughts and emotions welled up inside me. Often, a stream of everything I wanted to say to John, Gaie, or Laser would drone in my head while I harvested sweet potatoes. I could not shut it off, and with every new offense I grew angrier. Sometimes I could not hold back the tears.

The only other time I had experienced a world with few distractions was on the ship the R.V. *Heraclitus,* out in the middle of the Indian Ocean after twenty days at sea with only the sky, the waves, the dolphins, and flying fish beyond the microcosm of the boat. As I sat on the bow, my mind was free to shed some of the filters we all throw up to help us deal with the onslaught of daily life. Or, better stated, I began to see the shape of some of those filters: the prejudices, assumptions, hopes, and desires, which all changed the color of what I saw and the tone of my actions.

I began to see how my own English culture shaped how I thought of people. I am from a respectable English home, which meant form was everything when I was growing up. How I sat at table bolt upright with my elbows in, how I held my fork was reinforced day after day.

Even as a child, I was aware that form sometimes took precedence over practicality. I would watch my beloved father push peas onto the top of his fork, at least half of which would roll off the downward-sloping tines, as it was impolite to hold the fork like a shovel with the rounded part of the fork then able to cradle the peas. After I grew up, even an absurdly simple thing like how a person held a fork would change how I valued the person. But as I looked around the international crew on board the *Heraclitus,* I saw all manner of fork positions, and came to understand that what means good upbringing in one situation can be meaningless in another.

Here at Biosphere 2, I had been seeing my colleagues and my leaders through the filters of my love for our way of life prior to Closure and my devotion to the vision of Biosphere 2. I had filtered out what I now view as inconsistent behavior by John and Margret, in order to continue to build a glorious picture of the Synergia way of life and the project. My desire to be on the crew prevented me from seeing clearly many of the shortcomings that I now know were there all along. I excused or ignored them until they built to such proportions that I was forced to question fundamental premises.

I could now look at my life and see that, while I had struggled hard to get where I was, I was among the luckiest of people to have lived and traveled in places most only dream of. I had met and worked with legendary people. In my thirty years on planet Earth, I had already done more than most do in an entire life.

The bitter irony was that one of the people I had come to hate most in the world, John, had helped make it all possible.

And I saw that with all that, rather than running toward some lofty goal, I was running away as hard and as fast as I could. I had spent my life running away from myself, from that dreadful sense of a hollow core, a black hole of insatiable desire that nothing seems to satisfy.

But there was one person I could say was a truly dear friend for whom I would have done anything. My love for him grew stronger as the weeks and months passed. I clung to Taber and our love like a life raft, knowing that together we would float to safety, where alone I felt I would drown in the undertow of hatred and anger that filled Biosphere 2.

Also, we had Roy and Linda. The four of Us hung together, talking over the unhappy situation, and buoying each other up when someone was down.

Roy encouraged all kinds of extracurricular activities. He set up evenings with the Electronic Café in Santa Monica, which included jamming on the telephone with the musicians in the café. Once we played with musicians in Germany and Santa Monica in a musical round robin.

On July 20, 1992, Roy organized the Mud-wrestling Olympics in celebration of Taber's birthday. Using a drained rice paddy that we had just harvested, Taber and Roy performed a smackdown fight in the mud for the other six biospherians and an audience of staffers lined up at the window, clapping and cheering.

Dressed in nothing but baggy shorts, the two looked scrawny. Roy, the oldest on our crew, resembled an old rhinoceros, with his skin hanging loose on his torso. Taber, the youngest crew-member, looked like a skinny teenager who has outgrown his fat supply.

Gaie announced the start of the fight over her two-way radio and Roy leaped from the spaceframe, knocking Taber down in the gray mud. They rolled, they threw each other, they headlocked with great showmanship, until finally Taber flung Roy over his shoulders and splashed him facedown in the mud to claim his victory.

It was completely staged, but the crowd went wild as the victor, dripping with watery gray mud, waved and beamed. No one would ever have known that the eight of us were barely speaking.

In truth, the atmosphere had continued to be stormy after the SAC meeting on July 5. The research plan and its cover letter had only served to galvanize the SAC's resolve to oust John, and they were preparing a report on the science program to be released toward the end of July. All eight of us knew that the report was not going to be very positive and the tension was beginning to show. People arrived later and later to Agriculture crews.

The evening meals had been a good venue for everyone to stay abreast of the day's activities and discuss problem areas. Since we had abandoned them, none of us were in touch with what was going on in areas that were not our direct responsibility. We tried to have informal lunch meetings, but the first and only such meeting erupted in an *ad hominem* fight between Laser and Mark. The two became so loud and hostile that Linda, Taber, and I leaped to our feet and ran out of the meeting in disgust.

A couple of days after Taber's birthday, Taber and I butchered Sheena's kids and that evening we had a mouthwatering dinner of fresh liver, which I guzzled, although I rarely ate meat at the time. It tasted all the sweeter because that day the much-awaited report of the SAC had been released and the four of us saw a glimmer of hope that the beleaguered scientific credibility of the project could be saved.

The report still saw tremendous potential in Biosphere 2's scientific future, stating that the "committee is in agreement that the conception and construction of Biosphere 2 were acts of vision and courage. The scale of Biosphere 2 is unique and Biosphere 2 is already providing unexpected scientific results not possible through other means (notably the documented, unexpected decline in atmospheric oxygen levels.) Biosphere 2 will make important scientific contributions in the fields of biogeochemical cycling, the ecology of closed ecological systems, and restoration ecology."

However, as expected, it advised that John be replaced as director of research and development, infuriating the management and its allies. It was clear that John had lost this battle with the SAC.

But the war was not over. John had not lost his job, he was simply being moved to the side, so he could continue to provide the project's vision while a new scientific director would provide the professionalism the project required.

Everyone was angry at some aspect of the report.

When I reread the SAC report now, I see how naïve and short-sighted almost everyone was in their response to it. The recommended changes made a good deal of sense. It was an extraordinarily well-thought-out document. After extolling Biosphere 2's virtues, the committee then asked every difficult question that scientists would pose or were posing about the project regarding scientific validity, then answered each of them. For instance, the issue of there only being a single Biosphere 2 is a thorny one for most experimental scientists. However, they answered their own question as follows:

"The scientific method is often incorrectly understood to only involve observations of a system in which the parameters can be controlled by the experimenter. This is usually the case in laboratory experiments, in physics and chemistry especially. But in many sciences, this is not the way progress is made. Science may be observational and 'learning by doing' is a valid approach. For instance, in Biosphere 2 the unanticipated discovery of the decline in oxygen levels is an important result for which it is now necessary to determine its cause. Moreover, numerous scientific controls are in fact available to the Biosphere 2 scientific enterprise. On a small scale, comparisons can be made by varying conditions within the original, small test module or within other specially designed laboratory chambers. In addition, a series of comparisons can be made by successive Biosphere closures having different initial conditions.

"While a simple system might have been desirable for ease of scientific management, it would quite probably have precluded the kind of complex interactions that need to be observed and understood."

The report even outlined a research plan as the members would have approached it. Scientific success as they viewed it lay in organizing the research around six primary questions. They dealt with geochemical cycling over time, the interaction between the biomes, population genetic variation, biological-atmospheric-hydrological dynamics, and "the approach of these elements toward equilibrium in a materially closed system of a large scale." They wanted to answer the question of how one designs and monitors a nutrient cycle for humans with emphasis on agricultural production. And lastly they proposed we address the "relationship of biological diversity and ecosystem function in complex ecosystems."

The answers would "lead to a greater understanding of controlled closed systems and their design and function." They saw this point as the overarching scientific quest, which would also provide valuable information for conservation biology. The most notably missing question I would have added is, "What are the effects of living in a closed ecosystem on the human population?"

Gaie was hurt and angered beyond words. She thought we had answered all of the SAC's concerns during the many hours we had spent drafting and repeatedly revising the research plan. She took the recommendation for more credentialed scientists as a personal attack. Her only conclusion was that the SAC was out to get her, John, and the project.

Despite the fact that Ed Bass was apparently somewhat irked by the report, he released it for public review as promised. In his July 20 letter thanking all the committee members for having given so extensively without compensation, he tactfully wrote, "Though certain aspects of the report are frank and not necessarily flattering in their presentation of critique, its release ensures maximum candor and information to the public."

Communication between the four of Them and the four of Us was now limited to formal meetings, so we openly exchanged little about the report's conclusions. The captain's log, written daily by either Sierra or Laser, succinctly stated, "The Report of the Scientific Review Committee was issued to the media."

However, body language amply demonstrated Their state of mind. Gaie and Laser seemed particularly rigid with fury. They walked stiffly, their faces tight and drawn.

I had always served as the hairdresser in the group, and Laser stopped letting me cut his hair. It grew long, dark, and lank, which accentuated further his pale, bony, pinched features. He looked quite ill.

The personal strife was so searing that it was hard to remember what an accomplishment Biosphere 2 was, in spite of all our problems, or "challenges," as Margret liked to remind us to call them. We were almost a year into the two-year mission and we were still inside, still healthy. The oxygen decline had slowed, at least temporarily, because of the increased plant growth in the summer months, and we were still eating only from the Biosphere food stocks. All of the Wilderness biomes were still functioning well. The waste recycling and the water systems worked flawlessly, except for the odd freshwater flood from human error, messy but harmless. Despite the wrangling over the research plan, good science was being accomplished.

The architectural design had won the prestigious Gold Nugget Award from the Pacific Coast Builders Conference for excellence and innovation. Biosphere 2's architects, Phil Hawes, Margret Augustine, and the other twenty or so team members (for it was truly a team effort) had triumphed over entries from all countries around the Pacific Rim to win what was considered the Academy Awards for architects. Discovery Channel also named Biosphere 2 number three in a list of the twenty best architectural monuments of the twentieth century.

We should have been celebrating our accomplishments, patting ourselves on the back for having come so far. Instead, we were in purgatory. Yet in spite of it all, none of us would ever have dreamed of calling it a day, to cry uncle and be the first to beg to walk out the airlock door.

On July 24 we celebrated a bumper sweet potato harvest. Sierra made her signature sweet-potato pie. As if to remind ourselves that we were all decent human beings in an impossible situation, the eight of us spent the evening telling stories about how Biosphere 2 had come to be. It was a rare moment when we were all in agreement, and the atmosphere was even warm. Building Biosphere 2 had been an extraordinary accomplishment and we were all proud of it.

It was a touching time, and while the evening did nothing to mend bridges and bring the team back together again, it demonstrated that underneath all the agendas, every biospherian still had heart. It served to remind us why we were all imprisoned together, and that in a different time we had all cared for each other.

Chapter 16

STARVING, SUFFOCATING, AND GOING QUITE MAD

A s far as I could tell, we were responding adequately to the SAC's concerns. Their report had been stern in its indictment of the science at the project. In the cover letter to the media, Tom Lovejoy, the committee's chairman, wrote, "Although good potential exists for SBV to be a top-grade scientific effort, several related factors have held back scientific development. . . . We believe that given an adequate scientific reorientation, Biosphere 2 can fulfill its vision and become an important and unique contributor to scientific knowledge."

Harsh words indeed to a management who thought they had created the project of the century. The big question was whether the actions underway to respond to the SAC's concerns were simply window dressing, or whether John and Margret in particular, would work with the committee to undergo an organizational "scientific reorientation."

The SAC was not asking us to abandon biospherics. The advisors were simply attempting to make the project into a professionally run research organization that would be accepted into the scientific community. It was an acceptance the project desperately needed.

In the month after the release of the SAC Report, the crew spent many hours following the recommendations that we could. We set about writing several scientific papers for publication. Roy and

Gaie reworked the research plan, again. A new "import and export policy" was established. Rather than trying to process all scientific samples ourselves, we would send most of them out of the Biosphere to external laboratories. This would boost the amount of research we could accomplish inside.

Roy was particularly happy about this new policy. For some time, he had voiced his unpopular opinion that adherence to complete material closure for the two years was very limiting. It "was of no great scientific value in itself. In relation to Biosphere 2's educational and proprietary objectives, it did promote the corporate public image of a heroic venture, stimulated tourism and business interests, and perhaps increased the educational value," he wrote in his 1996 paper entitled *"Biospheric Medicine" as Viewed from the Two-Year First Closure of Biosphere 2* in the journal of *Aviation, Space, and Environmental Medicine.* The SAC had agreed and Roy was now getting his way.

We exchanged crates of soil-filled plastic baggies, bottles of urine, vials of frozen blood, plastic jars of Ocean water, "bombs" of air samples (an unfortunate nickname) for empty sample containers. Each biweekly import/export event was carefully orchestrated to trade the minimum amount of air with Biosphere 1—around two hundred cubic feet, or approximately 0.003 percent of the total volume of Biosphere 2.

We all loved these moments when the outside world reached into the Biosphere. They brought real, material contact with outside, even if the only things we touched were plastic bags filled with sample bottles, and not the warm hands of the person who had sent them in to us.

Like people living on submarines for six months at a time, no biospherian was expected to get sick with communicable diseases for the entire two years. We had all had each other's illnesses before coming to live inside Biosphere 2 and there were no new ones for us to catch. The exceptions: two bouts of flu sent in by someone on the outside who had some flu virus on his or her hands when loading sample vials into the airlock.

Gaie labored over research budgets and met hour after hour with Mission Control on the management structure discussed in the SAC report. Advertisements in *The New York Times* and elsewhere sought the new director of research.

On August 10, Mark, Linda, Taber, and Gaie all spoke via Pic-tureTel to a group of students in Japan at the International Space University, or ISU, as everyone calls it. At that time it was a nine-week summer session attended by international undergraduate and graduate students who wanted a solid grounding in all things space. Taber had attended the inaugural session in 1988, about which everyone involved was very nostalgic, and all ISU students seemed to feel as though they were members in an elite club. Now, one of their own was inside Biosphere 2. The students were very enthusiastic about chatting with us.

The session was held in a different place every summer and this year it was in Yokohama, Japan. Bernd Zabel, the original captain of our crew, was a student. On his return, he brought us his own ren-dition of Mount Fuji to look at through the visitors' window. Rather than the serene landscape for which it is so famous, Bernd painted Fuji erupting in violent flames.

It seemed an appropriate image for the current state of affairs at Biosphere 2.

In spite of a spectacular sweet potato harvest and an excellent peanut crop, we couldn't find enough food to feed the adult pigs. The domestic animals were supposed to eat what we could not, but the pigs were in direct competition with the human population. Their digestive system is similar to ours, so like us, they needed high-quality food, and could not be content with leaves and stems like the goats ate.

When the pigs had been included in the animal system years before, we expected to have an excess of food, which the pigs would eat. Unfortunately we never enjoyed a surplus. Now we took the painful decision to render the pig population extinct.

On August 28 Taber and I led Quincy into the small room at the entrance to the animal bay we used for butchering. He stood quietly while Laser put the stun gun to the center of his forehead and pulled the trigger. A pin with a round hammer on the end shot out of the barrel, inflicted a sharp blow to Quincy's head. He died instantly, collapsing onto the concrete floor.

As killing goes, it was humane.

Taber and Laser heaved Quincy's heavy carcass up and chained him by the hind legs. I cut his throat, catching the warm, red blood in a stainless steel bucket. Sierra would make blood sausage with it and the intestines that I washed meticulously. We wasted nothing.

Zazu followed a few days later.

I was sad to see Zazu and Quincy go. It felt like a betrayal to eat them. They had been with us for several years, and it was like eating a friend. The little of their meat I tried to eat was as tough as old boots. We were thankful for it, nonetheless.

On Sunday evening we toasted their transformation and their example of nutrient cycling in Biosphere 2. I wondered where their atoms would end up.

During the summer, we had not rained at all on the Savannah to keep it dry and dormant, which was part of its natural life cycle. It was now a wide expanse of untidy brown grass. In early September, as we neared the fall equinox, we planned to bring the Savannah out of dormancy to keep the CO_2 low. We harvested and stockpiled all the dead grass to store its carbon, and to make room for the new shoots to grow rapidly and unimpeded, soaking up carbon dioxide and putting badly needed oxygen into the atmosphere.

The oxygen was declining again and was now at just over 16 percent, 5 percentage points below normal. It was not yet dangerous, but we were all beginning to feel its effects.

For some reason no one understood, we were not adapting the

way mountain climbers do. Our bodies felt heavy. Performing simple daily tasks took huge effort. We could only move in slow motion, so much so that those on the outside who saw us regularly could tell the difference when we had our bimonthly cup of coffee—we actually moved at a normal pace for part of the day! Between low calories and reduced oxygen levels, harvesting the Savannah was a major chore.

The first anniversary of our mission, our grand experiment in isolation, was drawing near. We were again enjoying huge amounts of media attention. We all participated in TV interviews with NBC, CNN, European and Asian television, and satellite uplinks to TV stations around the world. Local schools came by to meet the bio-spherians and peer into our world. Cameras rolled. Anticipation and excitement were building—we were nearly halfway there.

After the bumper sweet potato harvest, we dug up plots of peanuts and planted sweet potatoes. The latter entailed sitting hour after hour on an overturned bucket, making ten-inch-long sweet potato plant cuttings that we then planted in the prepared field.

The field had already sat empty for too many days, so I sat all morning and all afternoon, cutting and cutting, for days. With every passing hour my backside hurt more and more, but I could not—I would not—stop, determined to get the damn thing finished.

When I got up on Saturday morning I bent down and could not get back up again. I had somehow strained my back and it froze.

Roy decided that the best way to speed up recovery so I could be back at work on Monday was to put me in mild traction over the weekend. I lay on my bed on the mezzanine of my room, with straps tied around my ankles, attached to gym weights that dangled over the mezzanine wall. I lay there all weekend, took Monday off, and was back at the potato patch on Tuesday morning.

During Tuesday lunch break, Sierra and Gaie walked into my room, and Margret called on the telephone, apparently with Nor-berto at her side. Without any small talk she announced, "I wanted these people to be there because I wanted them to hear what I am going to say. Harlequin, you should come out of the Biosphere. You've hurt your back, and in my experience a hurt back never really heals. I don't want you to feel that you have to lie about your back just so you can stay in your job; you can't do your job. Plus, I'm concerned about the insurance ramifications."

I was flabbergasted. She had offered nothing like, "I hope your back is doing better," or any kind word, just a big dump. And she had not even consulted with Roy, our doctor. Even Sierra and Gaie looked surprised.

I stammered something about waiting until Friday to come out, as I was aware that she was going to Fort Worth to meet with Ed the next day. I was fairly sure that Ed would not approve of sending a biospherian out for something so minor.

I never heard another word about leaving because of my back. Just a couple of weeks before our first-year anniversary celebration, Margret wanted to extract a biospherian for an injury she knew practically nothing about. I could only imagine the public-relations disaster.

Much of the crew felt unsupported by Mission Control and the management. That's not unusual. In recent years, Dr. Nick Kanas, professor of psychiatry at the University of California in San Francisco, surveyed fifty-four astronauts and cosmonauts who had flown in space or participated in a 135-day Mir simulation, and fifty-eight Mission Control personnel from the U.S. and Russia. The survey showed that over time, crewmembers felt less and less supported by their leader and by Mission Control. Astronaut John Blaha and others reported that during their four-month missions on the Russian Mir Space Station, they felt increased strain from working with key ground-support personnel who had little or no training in their position and thus provided marginal assistance.

To make matters worse, people in isolation tend to become

oversensitive to stimuli; how Mission Control communicates with a crew can have a large effect on morale. During his stay aboard *Salyut 7*, cosmonaut Valentin Lebedev wrote in his diary, "Every minute we have to keep ourselves in complete control. . . . Inappropriate words or jokes can put us off balance for an entire day." I agree.

The same survey found that at times of increased stress, crew displaced tensions to Mission Control and in turn, Mission Control displaced tensions to management. This displacement results in further disruption of interpersonal relations and can result in miscommunications. At the risk of oversimplifying, it seems to me now that as the difficulties piled up inside and out, we all started pointing up the management structure to find the blame.

Also, it is very common for crews in isolation to break into two factions, as we had. It happens in the Antarctic, in space, and just about anywhere small groups are shut off from the rest of the world. In the Biosphere, we were living out some preordained group madness.

Everyone seemed to be losing perspective. Mount Everests were made out of miniscule molehills, or so it seemed to me. Something as simple as a sharp tone of voice from someone in Mission Control during a videoconference was enough to ruin my day. I could not be sure that I was not losing perspective. Perhaps my back problem was a huge deal.

It seemed there were two projects running in parallel universes. There was the one where we had done the impossible—we had built Biosphere 2 and were about to celebrate a full year of closure with thousands of fans cheering us on. And there was the other project: tumbling down a vortex of infighting of which we were all guilty, which threatened to squelch the project's success.

Gaie and Laser telephonically joined a marine workshop in Mission Control to review research accomplished to-date and future goals, and to prepare for improvements to be made to the Biosphere 2 Ocean once the first two-year experiment was completed.

Dr. Phil Dustin, a marine biologist with the College of Charleston, and Julian Sprung, a world-renowned aquarist, were enthusiastic about the scientific potential of the Biosphere 2 Ocean. They were developing a long-term comparative study of the Biosphere 2 coral reef with an analog site in Belize.

They had already made one interesting observation. Earlier in the year, some of the corals in Biosphere 2 had started bleaching. It turned out to be a bacterial "white band disease" seen in nature also. What the researchers unexpectedly discovered was that it was not specific species that succumbed to the disease in Biosphere 2, but that some individuals had a greater tolerance for it than others. They confirmed this in the Biosphere 1 analog in Belize. This was good news for our hope of maintaining biodiversity on stressed reefs.

On September 22, we celebrated the fall equinox, which came with mixed emotions. We were happy to eat our fill and celebrate our recent harvests with a treat of peanut butter cookies.

But we all dreaded what the short days of winter might bring: high carbon dioxide, lower oxygen, and paltry rations.

I wish I could say that the awful parasite called self-pity never made an appearance in the Biosphere, but that would be a lie. Sometimes we wallowed in it. Self-absorption is part and parcel of being enclosed, I believe, and self-absorption is only a step away from self-pity.

Even so, excitement and anticipation had been building toward September 26, the first anniversary of our mission. The big day finally came and it was splendid, filled with drama and theatrics. Thousands of people came to celebrate. Each one of us took turns meeting and greeting people through the glass, answering questions about life on the inside.

Margret cut a huge birthday cake on the outside, while we cut a smaller, but still large, one on the inside. Sierra cut it into eight gigantic pieces and each biospherian gulped it down, while satiated people on the outside stared disbelieving at our appetite. The

Little Ugly Monkeys, a local Mariachi band made up entirely of teenagers, serenaded us. The celebration continued through the twenty-seventh, when Ed Bass came to the window to congratulate us on our efforts. We were—appropriately, it seems to me in looking back—terribly proud of ourselves.

Temple Granden, a world-renowned autistic veterinarian who has performed miracles working with commercial domestic-animal systems to make them more humane, visited. She toured around the Biosphere's perimeter with Safari and came across Mark Nelson at the visitors' window explaining to journalists about how the recycling system worked.

As she walked away she turned to Safari and exclaimed in her usual blunt manner, "Gee, the animals sure look good, but that Mark, he doesn't look like he is going to make it much longer!"

On September 28, we had our own private celebration. It was our biggest feast yet, with pork ribs from the freezer in sweet and sour sauce, rice with peanuts, stir-fry vegetables, potato chips, a garden salad, and chile bread, all followed by crepes stuffed with fruit, cake, sweet-potato pie, cheesecake, banana bread, papaya juice, and home brew. There was absolutely no way to eat it all, so everyone squirreled some away for later. What heaven!

But soon afterward, with the celebration over, the feasts digested, and the daily grind back, the feelings of inspiration and exhilaration gave way to a more ambiguous sense.

"Great, fabulous, wonderful," we thought. "We've made it through the first half of our mission but . . . oh no, we're only halfway through. Now we have to go through the whole damn annual cycle all over again!"

A few days after our celebration, I could hardly pull myself out of bed. I felt numb from head to foot, heavy, so heavy. All I could think was, "Just let me lay in this bed with the covers pulled up over my ears so I can't hear my chores calling me, and the pillow over my eyes so I can't see the Biosphere beckoning me."

The weight of my own body was crushing me. I was, in fact, clinically depressed.

Taber also succumbed to bouts of severe depression, as did others. Plenty of candidates for the cause of our melancholy existed in the Biosphere. Low oxygen contributes to mood swings and is known to cause depression. Low calories make one feel exhausted and the emotional burden of another winter on the way also contributed. We were all in a bad way.

Those who study people in isolation think the third quarter is the hardest quarter of any sojourn in confinement. Mark vehemently disagreed, saying that he thought it was an oversimplification, and that no "quarter" was harder or easier for him.

As for me, I was sure it would never end. The psychological pressures of being locked up, enclosed with only seven other people and few distractions, built to a boiling point. The psychological baggage that we all carry around with us but normally have the energy to suppress began to float to the surface.

Late one morning I was harvesting sweet-potato greens to feed the goats for lunch. All of a sudden I was a young child in some family scene where I felt uncomfortable. I could taste the emotions, see the expressions on the people's faces with acute detail. This was no ordinary memory. It had forced its way into my mind with such lucidity, it was as if I had walked through a time portal back to my childhood. Several biospherians reported these vivid flashbacks, and I had several.

I was sure that what happened in my memories must have actually occurred, but honestly, I will never know if they were simply a figment of my overtaxed mind—a waking dream.

You might say that we were undergoing what people pay thousands of dollars to a psychologist for—therapy—but I did not know this process, nor did I know how to handle all these images and new emotions coursing through me. I simply was not prepared to deal with it. Anger, crushing sadness, hatred, shame, confusion all burned into my heart, and imaginary arguments with one person after another forced themselves into my mind while I performed menial tasks.

At times I thought I was losing my mind.

It is very uncomfortable not knowing whether you are sane or not. Were we all mad, or was it just me?

Taber was certainly feeling similar turmoil, and as I looked at my other fellow inmates, I was sure that some of them were also quite ill. A couple of biospherians showed up for the early morning chores less and less often. They seemed listless and ate meals by themselves. Others seemed rigid with fury. Several had grown exceedingly thin and drawn. It was getting scary.

I have always been a pretty jolly soul—depression had not been in my vocabulary. I had always been taught to hold my head up, put my shoulders back, and get on with it. The English are a stoic, gung-ho lot, and when I was growing up we considered that only the seriously insane needed any type of psychological therapy.

But I needed professional help now. I did not know how to pull myself out of the bouts of depression, and I was afraid of losing myself in the mind-hammering flashbacks and the leaden blanket of negativity.

Now I know that there is a long list of maladies of the mind that astronauts and those who have been confined in the Antarctic experience, including depression and paranoia. A 110-day experiment, with three men and one woman enclosed in a Russian chamber in 1999, ended when a male tried to sexually assault the female. Cosmonauts have attacked each other with screwdrivers after several months of isolation in space.

One of the astronauts who stayed on Mir for four months returned a completely changed man. He declared it was the hardest thing he had ever done, and had he known what he was in for he would not have done it. He cried while he recounted his painful experience during a NASA debrief, all the while clutching tightly to his family.

Roy had been trying for some time to get Margret to arrange and pay for psychotherapy, but he had failed. She would have none of it.

I turned to Taber's mother, a therapist, for assistance. Doctors generally do not treat their own families. So Anthea found Taber and me a Jungian psychologist, Dr. Susan Schwartz in Phoenix, who agreed to have sessions with us separately over the phone. Initially I spoke with her twice a week. After a month or so, we talked weekly. She soon became my umbilical cord to sanity.

Susan was one of the few people with whom I could have an honest discussion who had no vested interest. She gave me badly needed perspective; she was my sounding board. She assured me I was quite sane, saying that she would have been concerned if I *weren't* undergoing some severe reactions to the events unfolding at the project and to the confinement.

Sometimes during these hour-long sessions, I felt like my insides were turning inside out. I had never been this exposed to anyone, even Taber.

There were three legs to the stool that held me up during this dark time: Susan with her reassuring, soft but firm voice proclaiming me quite sane; friends and family who suffered long phone calls full of venting; and Taber, who was a rock. The two of us would talk late into the night, airing our frustrations and struggling to understand what was going down. Taber, an eternally optimistic soul, would joke, "I promise these two years will end. Remember, there is life beyond Biosphere 2!"

I don't know how those biospherians coped who did not have someone intimate with whom to piss and moan.

But it was a schizophrenic society. On the one hand, we were all having extreme difficulty getting along with each other. On the other hand, as far as the world beyond the confines of the Biosphere 2 campus knew, we were all getting along famously.

However, when I asked later if people in Mission Control knew about the discord, Gary Hudman responded, "We knew. It was

between the Insiders and the Outsiders, and you guys were the Outsiders. Everybody you talked to had a certain take on it, and they
would always want to talk about it. Laser would call up and talk for
hours about it, ragging on everybody. Taber would always call up
and he loved to get everybody roused up. He particularly liked to
get Norberto going, and he was pretty good at it."

Nevertheless, we continued to pull together to get the chores
done, to maintain our Biosphere and, when the occasion called for
it, to party together. Occasionally a party would go sour, with one
member of the crew or another arriving in a foul mood and putting
a damper on the whole affair. But generally we all put our differences aside during these occasions.

On October 12 we celebrated Gaie's birthday with a rave in the
command room. We all danced on the granite conference table, and
performed strip shows, though as far as I recall, not one of us
stripped down to the full monty! Our spirits had been hoisted by
some fluorescent green ethyl alcohol Roy magically produced from
the medical facility. A friend had smuggled it past inspection at a
recent import/export by dropping food coloring in the alcohol and
labeling it "preservative."

People have often asked how we kept going despite the
upheavals. I think it was all the theater training that allowed us to
behave so differently in these various settings. Tony Burgess
recalls a conversation with Mark Nelson years ago at the Caravan
of Dreams. "Mark said, 'We do whatever we need to do, and play
what roles we need to play, to get done what we need to get done.'
That was the clue I got to the dynamics of your group. It was all theater. Play a part and play it well."

Some might call this deception. I call it liberating. One does not
get hung up on how credentialed a person is, but instead focuses on
how competent he or she is in the role.

I have heard too many times, "Oh I can't do that, I'm not an engineer." Or, "I'm not a writer." "I'm not a manager." "I'm not a . . ." Fill

in the blank. In the modern world we put so much stock in credentials and diplomas that they become a subconscious glass ceiling to
what we believe possible. In the world of our group, that was
stripped away.

We each had a role to play as captain, head of the Agriculture,
doctor, or analytical chemist. We each took on a character and
dressed in costume for birthdays.

Often the role-playing was conscious—I would walk into a
meeting donning my farm-manager role, and stay in character
through the meeting.

The trouble came when there was no specific character to hide
behind, when we were being ourselves.

Safari reported to me that things were worse among the Synergists
living together outside the Biosphere. Paranoia over the media poking
around in the group's private life blossomed uncontrollably. "We'd
normally eat dinner with the doors and windows open. But the paranoia got so bad that we had to install an air conditioner so we would
not suffocate because we closed all the doors and windows and curtains during Sunday night speeches. As we installed the air conditioner on a Saturday afternoon, Bernd and I just looked at each other.
We didn't say anything, but it was the exact opposite of what I thought
the Synergias represented. We were not supposed to demonstrate this
paranoid behavior."

People who had been with John for years were leaving. Others,
including Margret, showed up less and less to the formal dinners
and evening events. Safari and Bernd say Thursday-night philosophy had become excruciating.

Until this time, John had given seemingly well-thought-out presentations on a rainbow of ideas. But now, they say John seemed
totally unprepared and befuddled, pacing up and down in front of a
flip chart. "He was just rambling, saying that you guys inside were
the best people in the world, but us guys out here were all confused
and full of bullshit, and we should admire you. It was just an awful

rambling, with no content," Bernd says. "He was not even bashing us. It was just a rambling about the injustice. It was awful. So I left. After that, every time Johnny gave a presentation and he started snickering, and did big movements without an eloquent introduction, I left."

I had not witnessed a single person walk out of John's many lectures in all my ten years with this group.

The press was continually biting chunks out of our rears. *Buzzworm,* an environmental magazine, compared the project to an amusement park, and weighed in on the question of whether it was science or science fiction. A Hong Kong paper wrote about "Prisoners in a bubble critics want to burst."

Meanwhile, back in the real world, a group of eminent agricultural scientists gathered on October 17 to discuss the Biosphere 2 farm and its implications for tropical and developing world agriculture: Dr. Richard Harwood, Mott Chair of Sustainable Agriculture, Crop and Soil Sciences Department, Michigan State University; Dr. Jim Litsinger, an integrated pest management expert; and Dr. Will Getz from Winrock International, a nonprofit that develops and applies innovative agriculture systems in developing countries.

During the three-day meeting, Dick Harwood said that he thought our nontoxic, self-sustaining, intensive agriculture system was the wave of the future, combining the best of ancient systems with the best of modern agriculture. He validated our hard work, and inspired us to keep going.

A few days after this highly successful workshop, Sierra announced that she would be running the Agriculture crews thereafter.

I was devastated. Furious. Livid. I stomped around my room, wringing my hands, venting to Taber, and feeling completely victimized. There was no objective reason she had fired me as far as I could tell. No one had told me I was doing a bad job. It felt to me like Sierra wanted the glory now that the Agriculture had been declared a success by top scientists.

For some reason that I never understood, whether it was a sudden lapse into compassion, or whether she was overwhelmed by the prospect of running all the crews, she gave the Agriculture back to me, sort of. I ran all the crews, but sometimes she would dictate what they would be doing.

She became in charge of the water because I had flooded the basement one time too many. The irrigation water system was silly because it had to be operated manually with no shutoff valve if the tanks were getting too full. After she flooded the basement for the umpteenth time and I felt utterly self-righteous, we let rip a screaming session that left us both hoarse. I got the water systems back again.

And so it went on, she taking authority and then giving it back again, me feeling alternately victimized and righteous.

The four of Us all felt we had become the underclass in our dystopia, left out of any major decisions, and at times stripped of responsibility. Roy thought that Taber and I were particularly singled out, but Linda and Roy were not untouched.

John put Laser, who was no botanist, in charge of watering the terrestrial Wilderness biomes. "More water, more water," became his battle cry, convinced that the more he watered the plants, the more oxygen they would pour into the atmosphere.

Linda, who did know what she was doing, was convinced that this would simply put more carbon dioxide in the air from increased microbial activity in the soil and stressed plants that do not like waterlogged soils. It was nuts, but all she could do was stand by and watch.

On October 22 Norberto, now head of Mission Control, announced at a meeting between him, Gaie, Laser, Sierra, and Taber that the analytical laboratory was to be moved outside the Biosphere so that more science could be accomplished.

For several months Taber had been asking Margret and Norberto for help processing data from the technicians in the laboratory outside Biosphere 2. Taber was able to process all our samples, but

Later, we all sat in silence eating dinner. I tried to start a conversation about our chat with Jane, and was met with more silence. No one wanted to reveal any more about their inner lives to the other seven, although they had been eager to do so to someone on the outside.

Toward the end of her presentation, Jane made a statement that has stuck with me: "I have recently been at an environmental conference where eminent scientists were saying that we humans are on a train with no windshield wipers, no lights, and no brakes, going faster and faster down a dark tunnel. In other words, they were saying there is no hope for the future. I replied that humans are a problem-solving species and that with our intelligence we will learn how to fix the brakes and stop the train."

These were inspiring words indeed, full of optimism and hope, but I was not sure that the way things were going at Biosphere 2 set a good precedent. We did not seem to be able to fix or stop our train, but were instead wrecking it ourselves. In the end, it seems that the success of projects, the likelihood humanity will come into balance with the world's resources, is less about technical prowess and scientific adroitness, and more about whether we will use our intelligence, or will instead allow too much culture, psychology, and the seven deadly sins to get in our way.

Again and again, as I explore what went wrong during our two-year mission, I come up against the same answer—the technology was sound, the idea was noble, the project was visionary and courageous, but simple human frailty brought the walls crashing down.

I do not mean to suppose that all the blame lies with John, Margret, and their allies. Indeed, all of us were involved. It is simply easier to see the faults in others than in ourselves.

On Sunday night, November 8, I renounced my name Harlequin and regained Jane as my identifier. I thereby officially stated my separation from John and his cronies in every manner other than the employer/employee relationship.

The next day Roy, Linda, Taber, and I laughed until our sides hurt at a speech from Gaie. With absolute earnestness, Gaie had recounted an enlightening discussion she had had with John. Apparently John had decided that the Biosphere definitely proved Vernadsky's theory about biospheres operating on a geological scale because there must be a geological error in Biosphere 2. "There is not enough geology," he apparently announced to Gaie, "which is why the carbon dioxide and oxygen are acting up."

As the four of Us recalled her speech we chanted "Geologic error! Geologic error!" dancing up and down, and every now and then throwing in a gratuitous "More water, more water!" until tears streamed down our faces and we collapsed, exhausted from the black humor of it all.

The carbon dioxide had been steadily rising since its low in July. The cloudy days of the summer monsoons in August and September had brought it up to hover around a daily average of 1,000 ppm. Now the average was up at 2,000 ppm, with highs near 4,500 ppm. At such levels it again became a problem for the Ocean. Taber and Gaie monitored our little sea's pH closely and routinely poured in bicarbonate to bring its pH back up to just over 8. Taber turned the CO_2 scrubber on again.

Since the monsoons, the clouds had not really parted, and what meteorologists and climatologists had said was impossible had happened. We were in the middle of another El Niño-Southern Oscillation event. The clouds were thick and heavy. The sky was dark and condensation was dripping in the Habitat again.

The oxygen was still falling. It was now under 15 percent and we were still not adapting. We wheezed our way up the never-ending flights of stairs and tramped zombie-like from chore to chore. It was like the end of a heavy workout at the gym. My inner trainer would yell, "Just one more rep, come on, you can do it. Just ten more square feet of soil to turn. Don't be such a sissy. You can do it. Breathe, gasp, pant. What ever it takes. Get going. It's just a little burn. Mind over matter!"

Roy and Taber consulted with high-elevation specialists and analyzed our blood. They made us ride a stationary bicycle to failure. We huffed and wheezed, sounding like old people with emphysema, hanging half-dead over the handlebars when we could push no further, while they dispassionately monitored our heart rate and blood oxygen levels.

According to our blood chemistry and oxygen levels, we were in no danger, but the symptoms were becoming a real problem. We suffered from sleep apnea, where we stopped breathing for a few seconds, followed by a huge gasp of air that woke us up. This happened every few minutes.

Taber ran tubes all over the Habitat from the oxygen generator to people's rooms. When the symptoms became too troublesome, several of us slept with a small tube positioned under our noses and strapped to our heads, which provided us with pure oxygen so we could catch some sleep.

The most disconcerting symptom for those on the outside was our inability to complete a sentence without taking a breath in the middle. We must have sounded like we were gasping for air, about to take our last inhalation, when in fact we were simply breathing.

After thirteen months in Biosphere 2, we were starving, suffocating, and going quite mad.

Journal entry, November 16, 1992: "I saw a heron fly by the IAB—the vision of its power, graceful flight, and intention I will keep with me 'til our flight ends next year."

Then I got a real scare. My period was late. I was sure the worst had happened—that I was pregnant. I could not understand how, given that "we believe in family planning," as Margret would quip on TV when someone inevitably asked if there was sex inside. If I actually was pregnant, I would have to leave the Biosphere for sure, and what a scandal that would be. So far our personal lives had been left out of media reports of the project, much to all our relief.

But a pregnancy would change all that—it would give a license to pry. It seemed as though I held my breath for days.

Finally the wait was over and the familiar cycle started and I rinsed out my "keeper," a rubber cup we used instead of tampons, which were not allowed inside the Biosphere because of problems with disposal (they do not readily decompose). It turned out that our low body fat had disrupted most of the women's periods—something that many athletes have to contend with.

There was certainly lots and lots of sex in Biosphere 2—almost every organism was doing it. I had expected there would be all kinds of hanky-panky among the four men and four women. I expected to hear the pitter-patter of not-so-little feet along the wool carpet at night, surreptitiously slipping into another person's room. I figured that even those who were not paired up at the beginning would begin to find each other pretty tasty after a few months.

But I was wrong. The two couples who entered remained coupled (Gaie and Laser, and Taber and me), and the other four people remained celibate for the entire two years, as far as I can tell.

Gender was never an issue. There were no jealous lover's fights, no sex-starved manhandling of women, no handholding or necking in public. I never felt any vicious stabs in the back from women who weren't getting any, or at least none that I could attribute to sex or lack thereof.

We were all frightfully professional about it all. It simply was not discussed.

We were hungry, though—really hungry, now. Our most recent sweet potato harvest had been very bad, and Sierra and I had been unable to come up with replacements quick enough. We raided the Rainforest for green bananas. We ate plate-loads of sweet potato greens, which tasted like eating a mouthful of sour weeds. The morning porridge and the lunchtime soup became more and more watery.

So far, Sierra had done a heroic job of allotting food and not allowing us to eat ourselves out of house and home, even though we all whined and moaned about how hungry we were.

Even so, on November 16, Sierra announced at lunch that we would have to start eating excess seed and emergency grain stocks. If the harvests continued to be bad, we would likely have to import food.

Gaie coldly responded that two people would have to leave before food was brought in.

Nonetheless we began eating food that had not been grown inside Biosphere 2. I begged Sierra to tell the PR department so they could send out a press release about it. We had been hammered by the press every time we even gave the appearance of hiding something. I brought it up at the few lunch meetings we convened, but was told to shut up every time.

On November 23, Margret and Norberto convened a happy little meeting during which they reminded all the biospherians that they were not being forced to remain inside Biosphere 2, but were free to leave at any time if any one of us was unhappy or concerned about our health.

Finally I could not stand it. Dreading another scandal, on November 24, I called the head of our PR department, Chris Helms, and told him what was going on with our food supply.

He urged me to talk with Sierra about it again, which I did that day at lunch. Again she told me to go to hell. I told Chris, who told Margret that he would have to send out a press release saying we had begun eating the reserves.

I felt sick to my stomach. I knew full well that if Margret realized I had told Chris, I would be fired for real this time.

At three that afternoon Margret called a meeting where she accused Linda and Taber of speaking with Chris. She was clearly on a fishing expedition.

I did not need science to tell me, but scientists have proven that "a stress response occurs in an individual member of a cohesive group who deviates from the beliefs or accepted actions of that group." Was I ever stressed over this episode of "deviant" behavior. However, the same scientists also showed that an individual acting against his or her principles in conformity with the group also showed stress. I had greater abhorrence for continuing to do something I felt strongly was wrong and damaging than for the inevitable punishment I would receive.

At 5:00 p.m. I was called into Sierra's room, where Margret accused me of going to the media, which it never occurred to me to deny. "The media?" I thought to myself. "Chris is one of your own. Your employee. He's on your side, trying to do right by the project."

In that moment Margret revealed that she thought of him as an outsider, an adversary. What I was not aware of was that Margret, John, and Chris had been going at each other that afternoon in Chris' office.

According to Chris, Margret screamed at him to tell her who had told him. He refused. John was playing the good cop, but after Margret shrieked at him to shut up, Chris says John curled up in a fetal position on the sofa. Chris walked out, threatening to resign if she kept up this intolerable behavior. Sometime later she called him and invited him to her office, which Chris agreed to on the proviso that she behave "like an adult."

Margret's response to this episode was to throw up more barriers to communication. She told Chris he had to go through Norberto now, if he wanted to know about anything going on inside the Biosphere. Chris, incensed, told her that that would not work because he had to have unfettered access to what was really going on inside, and that he quit. Finally she acquiesced, and Chris walked quietly away to write a press release.

The next morning Sierra and Laser held a poisonous conference with me, telling me that I would be leaving on December 15. I had one day to make up some story as to why I was out.

"Of course, you can go to the press with whatever story you want.

But if I were you," hissed Sierra, "I would leave honorably, and say that you have some medical problem, such as pregnancy."

All I could think was, "I went to our own PR department, bitch. The way you tell it, I went live on CNN."

Gaie and I had a yelling match. She clearly did not hold me responsible for my actions. She was just as convinced that I was under the spell of Linda, Taber, and Roy as I was convinced that she was John's little puppet. Gaie was on my side in an odd sort of way. She told me that she thought it wrong that I be sent out and that she would go to bat for me.

I knew that I had some choice in the matter of staying or leaving. I could have gone to Margret and begged forgiveness, offered a *mea culpa* and promised to be better.

I knew the workings of this dysfunctional family all too well. People were fired and unfired so often that it was almost as if John and Margret considered firing merely an extreme form of ordering someone to go stand in the corner. Margret had pushed me too far this time, and the question in my mind was did I really want to stay? What was the use in all this? Nonetheless, I also knew I could not live with myself unless I tried my hardest to complete the two years.

I could not sort out my thoughts and feelings, so I called a colleague for some perspective–Tony Burgess, the Desert biome design captain. Tony had been a long-time friend of the project and I hoped I could trust him not to divulge my conversation. In the course of our discussion Tony told me that he thought John had built his own world to rule, as he could not rule Biosphere 1. But I was dismayed when Tony told me that I should go to the press and really shake the tree.

I did not know how significant this phone call had been until twelve years later when Tony said, "You told me that John Allen had said you couldn't tell anybody that you were eating into the seed stock. That says coverup to me.

Tony's actions in response to this awakening would leave us all shellshocked.

Finally I talked with the people who had started this fracas—the guys in the PR department. Chris and his assistant, Scott McMullen, insisted that if I came out it could mean the end of the Biosphere. They did not know how they could possibly deal with the public relations nightmare, although, just as Chris had anticipated, there was no backlash from the press about our eating from the reserves.

The next morning, after I dialed in the codes, the digital woman announced a voice message: "November the twenty fifth, six seventeen a.m." My breath caught in my throat as I heard John's voice, which I had not heard for many months now. He was speaking tenderly, in a tone I had never heard pass his lips. "Hi Jane, this is Johnny. I don't really know what is happening. Please do consider staying on there. I think it would be a wonderful thing. You may have made mistakes. I've made lots of mistakes. We have all made mistakes. I think you are a good part of the whole thing."

His voice was becoming raspy, thin even. "Look, I know that you, Taber, and Linda don't like me very much anymore, and if I'm part of your decision you won't have to put up with me much more, Harlequin, or Jane. I am resigning as director of research and development. We are going to get a guy in from the system. Don't factor me in. I'm not going to be running the R and D for the last few months you guys are in there."

His voice was beginning to crack. "I don't know, Harlequin, I love you, I love all of you guys. I just screwed up a lot. I don't know what happened. I really don't. But believe me. You will never regret finishing the whole thing. You never will. And I know . . ."

His voice trails into sobs. "Anyway forgive me for anything I've done, Jane. . . . I am so sorry. . . . Anyway, don't react to anything you feel against me, because I feel really terrible."

His sobbing was echoed by my own now. "I am so sorry," he went on, "I just apologize . . ."

The message ends abruptly midsentence as the allowable message length runs out. I was left holding the receiver dumbfounded.

The atonal female voice broke in, "Message two."

It was John again, the pain in his voice streaming down the phone. "I feel total remorse for any contribution that I have made to this scene. I am horrified at the part that my failures played, but don't, don't add yours to mine. Don't add yours to mine, Jane. Anyway, I love you no matter what you do, I want you to know that." He finished by telling me he was off to Mexico for two days. I would call him when he returned.

I was thunderstruck, sobbing as I replaced the receiver back on the phone, my hand trembling. I had never heard such vulnerability from John. Maybe there was a chance we could patch all this insanity up and be all the better for it.

On November 26, we ate ourselves silly for Thanksgiving. Mildred, the piglet, was served on a spit. Roy captured me on film drooling over the piglet, saying, "Mildred, you never looked so good!"

It was amazing that we kept up these celebrations. I was in the midst of being fired, and each half of the biospherian crew thought the other half traitors.

Two days later, on November 28, Margret called me and was very charming. She quietly named all my failings, including that I behaved like a housewife for Taber, which was supposed to be an insult.

Nonetheless, I was to stay on in the Biosphere. I apologized to Sierra, who was icy. Laser and I hugged and cried. As we wept, we wondered how our relationship had become so acrimonious.

The whole incident could have been cathartic. That evening, all eight of us engaged in the Native American tradition of the talking stick. The person holding the stick speaks without interruption until he or she passes the stick along to the next person. We made a pact that nothing would be relayed outside the circle of eight. For the first time there was some emotional honesty. But it was not to last long.

Saturday morning I spoke with John in person. I could tell by the harsh tone in his voice when he answered the phone that this was not to be the conversation I had hoped it would be. As he droned on about my failings my heart sank—it was as if the phone message had not happened and the crew's moment of openness and honesty was sequestered safely between two parentheses of time.

I was shattered. John had laid me wide open with his teary phone messages. I had been ready to forgive almost anything to make amends if it meant a deeper friendship, and an honest assessment of where things had and were still going wrong. But now he had slammed the door on that possibility.

Time had taken on a peculiar quality—it seemed to several of us that we had never been anywhere but inside Biosphere 2, and that we would not be anywhere else ever again.

The Biosphere, *our* Biosphere, was our world, and time, along with our emotional roller coaster, was also sealed inside. I felt like the two years were never going to end.

Chapter 17

I DREAMED
OF WILD HORSES

O
utside, rumors abounded. We were sneaking out to bars at night. Children were born inside. A local paper even went to one of the nearest bars to ask if we had been spotted there. The barman's response: "If they were going to sneak out to a bar, do you really think they would come here?"

We were apparently also sneaking pizzas in. "If only we had!" I thought to myself. One Sunday morning, I went down to the visitors' window next to the airlock and saw a pizza box sitting neatly outside the door. While I knew it was physically impossible to sneak anything in or out without someone in Mission Control knowing (there were numbered lead seals on the door that had to be broken in order to open the door), I went along with the prank.

"Open it, will you?" I asked my visitor. She lifted the lid and there was half a pizza left. "No way it came from in here." I said. "We would never have left even a mouthful of pizza in that box." Case closed.

On November 30, Taber said goodbye to the ion chromatograph. A while later the atomic absorption spectrophotometer was carried into the airlock. It was a belly blow for Taber every time another piece of equipment left. So far, little that had been sent out had been put to use in the laboratory outside—the technicians couldn't get it functioning. What kept Taber going was his work on the oxygen loss, and the medical research he was doing with Roy.

Some of the biospherians and Mission Control had been discussing whether to add oxygen to our world, but each time they
decided to let it go a little further. As uncomfortable as the situation was, Roy and his consultants who specialized in high altitude
medicine assured us we were not in any danger.

On December 7, several of us started on Diamox, a medication
that helps alleviate the symptoms of altitude sickness. The medication made me feel considerably better, with more energy to work
and cope.

Prior to Closure, Roy and several consulting physicians from the
University of Arizona had concluded that the primary medical
issues during the two years would be infections and allergies from
the hot, moist environment and the abundant grasses; trauma from
living and working with so much dangerous equipment; toxicity
from material outgassing and from being in a materially enclosed
atmosphere; and psychological problems.

With the exception of my finger amputation there had been
almost no incidences of any of these problems. Although I felt I was
going insane at times, none of us suffered from any major
pathology such as psychosis, as cosmonauts have experienced in
space.

Instead of dealing with all the anticipated medical crises, Roy
had to launch research programs on completely unexpected medical concerns such as the low-calorie diet and our apparent mal-
adaption to the declining oxygen.

Roy and Taber closely monitored our response to the lowering
oxygen. Using the same instrument used in hospitals to measure a
patient's blood oxygen saturation, they clipped a pulse oximeter to
each biospherian's index finger to measure our blood oxygen saturation at rest. Then we climbed aboard a stationary bicycle and
measured it again while we pedaled as hard as we could until we
were panting, which did not take very long. Taber rigged a data-
recording device so he could measure each biospherian's blood

oxygen during the night, important because of the apnea some of us were suffering from.

What they found was that at rest and during the night, our blood generally remained saturated—normal—or slightly below, but did not fall to a worrisome point. But our blood did lose oxygen under hard exercise, making us feel weak or out of breath when we exerted ourselves during our daily tasks.

Finally, Taber drew arterial blood to measure the oxygen in our blood at the point at which it has the highest oxygen content, having just left the lungs. There are always risks inherent to puncturing an artery, so Taber chose only two biospherian wrists to stick—Roy and Linda, who were suffering the worst symptoms from breathing low oxygen.

The arterial blood was also saturated with oxygen.

Despite the obnoxious symptoms we were all feeling, the tests indicated that our bodies were handling the Biosphere's loss of oxygen. We all agreed to continue letting the oxygen drop to give us more time to discover the cause of the oxygen loss and why we were not adapting. Injecting oxygen might well mask both causes.

So down it went. The level was now dipping below 15 percent, a full 6 percentage points below normal. We did not know how low we could go before it became dangerous. Mountaineers occasionally die from elevation sickness when the person does not adapt to the paltry amount of oxygen in the thin air. It was highly unlikely that we would suffer such a devastating response because our oxygen loss had been gradual over many months, not in a couple of weeks, as mountaineers experience. Nonetheless, we all felt crummy.

TC, Biosphere 2's chief architect, paid us a visit in early December, almost two years after he quit. As we chatted at the visitors' window he sounded the most upbeat I had heard him in months. "I'm free!" he cried.

That night, and many nights thereafter I dreamed of wild horses, large herds of mustangs galloping across an open plain. It was a classic image of physical power, vitality, and freedom. I was one of them, with no bars, no glass walls, and no suffocating atmosphere.

It did not matter where we were galloping. We ran because we could, for the pure joy of it, an expression of our free spirit that welled up from deep inside. And I dreamed of flying, high above cities and wide-open landscapes, hurtling over crowded streets.

On December 9, the four of Us watched an eclipse of the moon while eight fat deer grazed on the landscaping below the Biosphere.

I realized that I was hearing few coquis. Where had the little frogs gone? In our attempt to manage our atmosphere by cutting down the Savannah grasses, in which the coquis lay their eggs, we had reduced their habitat and thereby their population. They were innocent victims of our environmental policies.

December 21 saw all eight of us celebrating solstice together on the inside. The group on the outside had disintegrated to such an extent that for the first time in many, many years, they did not celebrate it.

However, inside Biosphere 2, we started at lunchtime with a feast similar to Thanksgiving. After gorging, we all lay around listening to oldies all afternoon, and then invented the biospherian steam bath.

Laser came up with the steam machine that I think was supposed to be used for cleaning carpets and upholstery but had been left untouched in the mechanical room. We made a tipi of towels and blankets and all snuggled inside. After the sweat lodge of sorts, we all sat and watched the *Addams Family* on TV.

We had a genuinely good time together, in part because our mood had been mellowed by some booze that had miraculously appeared. If only it could have always been like that.

There was a reason that the Russians sent bottles of cognac to the cosmonauts on Mir for celebrations, and turned a blind eye to the vodka smuggled up in bags marked "water." Stimulants, such as coffee, also helped our state of mind.

The gaiety resumed a couple of days later when Roy hatched his "Butt-wheeled Wagon." In the plaza outside the Agriculture, seven of the eight of us smeared nontoxic body paint on each other. Roy's plan was to paint a huge mural on the blank white wall by using our bodies as the paint applicators. We painted our backsides and the backs of our legs with blue paint and then pressed each other up to the wall to make four spokes of two wheels. The men's butts formed the back wheel, symbolizing physical power, and the female bottoms made the front one, symbolizing new directions. The red hands of men and women joined to connect the two wheels. The green outline of two lovers—Taber and me—pulled the wagon.

There we were, stark naked in the plaza, hidden from visitors' prying eyes, wearing paint for clothes, all the while being filmed by Roy for some other art venture he was cooking up.

Christmas Day was quiet, with a small feast in the decorated dining room. The next day we found that Sheena the goat was very lame. At one of the animal bay's windows, a crowd gathered around Dr. Page, the vet, as she told us how to treat a deep infection in Sheena's leg. Someone who seemed quite plastered was heckling from the back of the crowd.

Taber and I doctored the goat daily. The whole procedure was so painful for Sheena that before touching her leg Taber jabbed her with a sedative straight into the jugular. I stood by to catch her, as the response was almost instantaneous. Down she went, lying like a corpse while we treated her leg, and about fifteen minutes later she was up and around again. After a few days of this it seemed that she was getting hooked on the stuff. When she saw Taber and me walk into her pen together she climbed up onto the box we usually injected her on, and offered her neck to Taber.

Biosphere 2's first junkie!

Finally, Sheena's leg healed and she was off the drugs.

New Year's Eve was another happy event—poker in the library, with peanuts as the chips. Actually, we have always said they were peanuts, but (here I divulge one of our dirty little secrets) it was a truly high stakes game as the chips were . . . M&Ms, a heavenly gift smuggled in from outside.

Taber walked away with the lot.

Sympathetic friends in Mission Control had become very good at hiding the odd "gift" in equipment that was imported to the Biosphere, building false bottoms in boxes, for example. We were all extremely serious about the integrity of our mission, and the few goodies we devoured on special occasions made us feel naughty, but not like frauds. The amount we increased our caloric intake with these few acts of hedonism was wholly insignificant. But the increase in morale was major.

Some members of the second crew, who would inhabit the Biosphere for just over six months, were more creative with their excursions. Matt Finn, whose primary responsibility would be the Marine biomes inside the Biosphere, convinced Gaie that he needed to fill large tanks with clams to help filter the Ocean water during the second mission. In fact, he had a different use in mind— the crew gobbled the lot during their stay.

Several crewmembers also undertook late-night stealth missions. The outer wall of the large airlock by the Savannah had a water trap to prevent over- or under-pressurization of the airlock. It was a one-inch U-shaped pipe that punctured the outer skin of the airlock. Water in the U sealed the pipe from air (the water had a film of oil to stop it from evaporating).

On one occasion, the wayward biospherians siphoned a bottle of rum and some beer through the water trap using a piece of tubing from the lab, while the accomplice on the outside shook in his boots, hoping he was not going to get caught red-handed.

Rodrigo Romo, another crewmember in the second mission, recounts a late-night prank with several of his crewmates involving a sealed pipe that punctured the Biosphere's skin next to the visitor's window. "One night Ben Jessop (an SBV staff member) comes down with a twelve-pack of beer, and we had posts on lookout. I was in the viewing room with Matt Finn, Charlotte was in the Rainforest, and Smitty was in the IAB. Anyone seeing someone approaching would say the code word *eggcup* over the radio.

"So, Ben Jessop was on the outside unscrewing the cap. Matt and I were on the inside unscrewing the lid, and it was the perfect size for a beer can to go through. So, Ben is sliding the cans through, and suddenly Charlotte calls on the radio, 'Eggcup, eggcup!' Ben nearly soils his pants. He went white and froze. He couldn't run or hide, as there was nowhere to go. And Charlotte is calling, 'Eggcup, eggcup, eggcup!'

"Finally she comes on and says, 'Oh, no, false alarm.' So poor Ben Jessop has almost had a heart attack. I say, 'He's gone man, come on.'

"But Ben's crying, 'No, no, no, no!' Finally we convince him to pass the rest through."

Despite the continuing dramas inside and out, the holidays had been a respite from the turmoil. That we had been sincerely happy during those days, apparently because of the added calories and interjection of the odd alcoholic beverage, left me wondering if our strife inside had little to do with the craziness that swirled around the project, but was more a product of our own psychological and physical discomfort. Would that it could have remained like that.

Chapter 18

DYSFUNCTIONAL FAMILY

E arly January 1993: Roy was not looking good, nor did he feel well. He thought it was from the low oxygen. His face was drawn and he moved even more slowly than the rest of us. We all ribbed him about a peculiar, tottering gait he had developed. He laughed about it, too.

None of us knew then that this was likely the first sign of Lou Gehrig's Disease, or ALS, which would wrack his body only a few years later, and ultimately kill him in 2004.

His resting pulse was sometimes almost double his usual heart rate, 110 beats per minute, and he was feeling foggy-headed. The oxygen was almost down to 14 percent, beyond the point where pilots and air passengers go on oxygen.

Although we had thought we were not adapting, Roy had recently discovered that we were, but in a way not normally seen in humans. As oxygen levels decrease, a compound in our blood that controls how much oxygen is released into the tissues usually increases. A rise in the compound allows more oxygen to saturate the tissues. An increase in red blood cells and return to normal levels of the compound generally follows. Such an adaptation to high altitude occurs over the course of weeks.

In Biosphere 2, the slow oxygen loss produced an oxygen decline (technically, the oxygen partial pressure) equivalent to an eighteen-month climb to fifteen thousand feet. None of us exhibited any significant increase in red blood cell count or total hemoglobin, and the compound had dropped in our blood, not risen, making the hemoglobin affinity for oxygen higher than normal, not lower. This allowed more oxygen to be in our blood than if the compound had remained normal, but it did not mean that the oxygen was reaching our gasping cells.

At that time Roy was not sure if the unusual response was because the level had sunk so slowly. However, he later hypothesized that, because we were all low on caloric energy, the crewmembers' bodies conserved energy by not increasing production of red blood cells. Instead, to respond adequately, our bodies manipulated the compound's level in our blood to boost the amount of oxygen in our blood. Since the life span of a red blood cell is limited, it is energetically expensive to increase and maintain a higher red blood cell count. This response is similar to that seen in hibernating animals, which have to deal with food deprivation and low oxygen over a winter.

No wonder we felt bad—our bodies thought we should be curled up in a cave hibernating for several months.

The thin air hit Roy particularly hard. He was concerned that his judgment was too impaired for him to respond to an emergency. On January 9, he asked Mission Control to add oxygen into the Biosphere. Mission Control told him that it would only be added in the case of a medical emergency, which Roy took to mean when someone was in danger of death within two days. He was outraged, but was in no state to push the subject very seriously.

On January 11, Taber walked into the clinic and found Roy hunched over his notebook, looking confused. He had been trying to add up a simple list of numbers and couldn't. It was a sure sign of the effects of hypoxia, and he made Taber the acting medical officer in his stead.

Taber had been trained to be the assistant medical officer, and to take over from Roy in a limited capacity in the unlikely event that Roy was incapacitated. He had undergone a customized medical training program at the University of Arizona's Medical Center, which included two weeks in the emergency room, a slew of training courses, and a Navy dental-school course that had come in handy when Roy needed a cap replaced.

While he had fully anticipated assisting Roy with the bimonthly biospherian health checks and with medical research, no one really expected that he would be taking over from Roy. But Taber launched himself into the oxygen debacle.

After many debates, Taber, members of the SAC, Mission Control, and external advising doctors finally reached a consensus that we had to have more oxygen. On January 13, with the oxygen level in Biosphere 2 at 14.15 percent, we walked slowly into the West Lung, closed the door and waited for the oxygen level to climb above an ambient level of 21 to 26 percent, at which time it would be released into the rest of the Biosphere.

As the oxygen rose in the lung it was as if an enormous weight had been lifted. A sense of euphoria overtook us and we began running and jumping, something we had not done since shortly after Closure. We were elated.

But, as we left the lung to go back to work, I felt like a ton of weight was added to my shoulders every few feet. By the time I walked upstairs to the kitchen I felt like what I imagined a frail 110-year-old must feel like—hardly able to move.

Over the next nineteen days, thirty-one thousand pounds of liquid oxygen were injected via the lung, bringing the oxygen level in the whole Biosphere up to 19 percent.

The ghastly sleep apnea disappeared and I could finally enjoy a full night's sleep. I felt resuscitated.

The media had a mixed response to the news, though very few condemned the action. Kathy Dyhr, who had been our PR person prior to Closure but was no longer with us, asked Bill Broad of the *New York Times* why the press had not pounced on us. He told Kathy that since we were upfront about the oxygen thing, there was no story.

On Sunday, January 18, already feeling much revived from increased oxygen, Gaie, Mark, and Roy sat at the command-room table participating via video-link in a meeting in the Mission Control conference room. It was the second day of a three-day SAC meeting.

The rest of us watched from time to time between our chores, or came in to give presentations to the illustrious gathering about the various science projects in the Biosphere—the oxygen loss and medical studies, Ocean, Marsh, and terrestrial biome research. Our command room felt more tense than normal. I assumed it was the usual stress of giving presentations on complex issues.

I was not aware of the undercurrents coursing through that meeting. Tensions had been building between the management and several members of the committee during the past few months. Despite the promise to replace John with a credentialed scientist as head of research, he was still on the job and clearly still in charge. The committee members were pushing hard for new scientific leadership. The management was incensed; they felt the SAC was acting outside its purview. They were convinced that some SAC members were attempting a coup to take over the project.

John had recently been calling the SAC the "establishment hit squad come to take over, headed by Ed Bass." I did not know what had caused the rift between him and Ed and the SAC, but I just knew this was not going to be good.

The discussions moved to procedures and policies for data quality control. Norberto, as head of the Mission Control team, gave a nebulous speech about the computer system, touting its novelty and efficacy. Later, he pointed to a five-foot-long row of loose-leaf binders stuffed with papers and concluded,

"If anyone wants to get into it in more detail, then you can read through this five feet of documentation on the Nerve System!"

I wondered why he was being so defensive.

The SAC wanted to know more about how information reached

the SAC, the science community, and the public. Tom Lovejoy asked, "What's the policy for who has access to what data?"

Gerry Soffen weighed in, "It appears random as to who gets the data and what is released to the press and public, perhaps because the policy has not been disseminated, or perhaps because it needs to be made. Policies are really good at defusing the outside community that is confused."

Ed responded, "Given that there is an enormous investment, and it is an entirely private investment in this project, the gratuitous dissemination of what might be considered valuable data is simply not going to occur. The intention is to disseminate all that is useful to humankind, but is not useful for commercial purposes."

Tom interjected, "I am not asking for a data cloud over Tucson." Everyone laughed, "But at the moment, the process of making those decisions is fairly *ad hoc*."

Ed calmly replied, "No, it's just not clear, because you have not seen a stated policy."

Finally, Margret ended the discussion by saying she would get them a one-page policy.

It was then John's turn to talk about the research goals. He wore the sardonic grin he got when he felt backed into a corner and was about to lash out. According to him, the first experiment—our time inside—was the maiden voyage, where the main objective was making the Biosphere work. After tightening a few bolts during the transition, the next experiment would be more of the same, and in the third experiment other research could begin. That was the total of John's research plan—the plan the SAC had asked repeatedly for over the past sixteen months.

Gerry, not surprisingly, wanted clarification, saying, "If I have understood what you have said, this first mission is really a test program. And you are saying that we will repeat the test program but with improvements. And the experimental program will really begin on the third mission."

John responded, "In the first mission, the scientific question is whether or not an apparatus like this would even work."

"But it's basically an engineering effort. You can split hairs on what is science and what is engineering, but your first goal is to make it work? The question I really have is: What do you expect to use the Scientific Advisory Committee for? I fully understand that you need to make it work, but I don't understand why you need a scientific committee. . . ."

John threw himself back in his chair, and with an awkward laugh quipped, "Well, that's a good question, as far as I'm concerned."

The room fell silent.

I was flabbergasted. He had insulted every member of the committee after they had risked their reputations, and had spent countless hours giving media interviews and in meetings. They had bestowed their credibility, their expertise and their experience on the project—*gratis.*

But the appalling scene didn't stop there. Steve O'Brien, a geneticist with the National Cancer Research Institute, chimed in, "It seems to me that the Science Advisory Committee was called together for damage control. I think that we have accomplished that because the publicity in the last six months has been pretty good. So if that is all you view this committee as, we have succeeded."

Gaie, Mark, and Norberto began talking at once, saying how wondrous the SAC members were.

A new member I did not know attempted to calm the waters, saying, "My understanding is that there are two goals, and the secondary goal is to squeeze out science while doing the primary goal (of making it work), which is happening."

Mark agreed, "Although our schedule is very busy, your advice on the research that we can make time for is very important."

Finally John apologized for his outburst, and Tom was conciliatory. A fly on the wall would have thought they had kissed and made up.

On Monday, the meetings were closed, so we all got on with maintaining the Biosphere. On Tuesday morning I was putting fodder in the animal bay when Gaie walked over to me.

I watched as she collected a big wad of saliva in her mouth and spat in my face. She turned and walked away without a word.

Forty-five minutes later I was walking up the spiral stairs during break time when Laser stopped on the stairs next to me and spat a mixture of saliva and peanut goo in my face.

"What have I done?" I asked, wiping the dripping, slimy mess from my face.

"That's for you to find out," came the surly reply.

Finally I discovered that they thought I had told the SAC about my being fired over the eating seed-stock affair. I had done no such thing.

To my horror, I realized the unintended consequences of my call to Tony Burgess for advice during the firing episode. He told the SAC during Monday's session, as he was concerned that management was trying to cover up food shortages.

Somehow news of Tony's meeting with the SAC reached the ears of some of the crew. The evening after the meeting Laser was beside himself about it, and he hurled a bowl violently to the floor. It bounced and hit Sierra.

Rumors had coursed through the crew about letters he and Gaie wrote to the SAC, stating their disapproval of the committee's actions. They said the SAC had been meddling in management and organizational structure, which they felt was none of the committee's business.

Laser was furious at Roy for not writing a similar letter. He also lashed out at Linda and me, claiming that we had told the SAC members all kinds of terrible things, which I certainly had not. He just had to know if we were lying, so he demanded a meeting with Tony.

On Friday poor Tony was put on the stand and cross-examined by Laser and several other biospherians.

Finally Tony said, "I think this whole scene at Biosphere 2 is analogous to a dysfunctional family."

"Bullshit!" fired Laser, slamming his fist on the table. All four of Them were greatly offended by the statement, though it seemed to me that Tony had hit the nail on the head.

At that moment I made a pact with myself. I was going to do everything in my power to stop getting sucked into the sickening group dynamics and make an effort to enjoy and experience Biosphere 2 for the last eight months of our enclosure.

Unfortunately, some things were out of our control.

On February 5, Norberto wrote a memo to all biospherians claiming that Tony had said that one or more biospherians were insane. Management had no choice but to bring in a psychiatrist to assess everyone.

"Finally," I thought to myself, "Psychology is being taken seriously."

Dr. Alan Galenberg, the head of the Department of Psychiatry at the University of Arizona, was a charming, soft-spoken man. He interviewed all eight of us privately to ascertain whether we were nuts.

He talked with Taber last, congratulated him, and said that he admired the four of Us for being able to use the situation as a transformative experience. He said he would recommend to management that they open an anonymous fund for psychiatric help. His report would also lay to rest any concerns about any of us being clinically insane, whatever that really means.

At the close of my conversation with him he said, "I had not previously understood how the food systems, the operations, and the research of the Biosphere are all so intermingled with who is who, in what position and how they react and interact with everyone else."

"Oh," I said, "You mean a dysfunctional family."

"Those were the words going through my head, yes—a dysfunctional family."

Chapter 19

WINDOW-SHOPPING

S o, what more could possibly go wrong in our self-inflicted mini-hell, our own peculiar rendition of *Lord of the Flies?* Valentine's Day brought a nasty eye-opener. Linda received a phone call from a close friend in Oregon saying that her local paper had reported that the Science Advisory Committee had quit.

The four of Us turned on the TV, and sure enough it was all over CNN. They even said Biosphere 2 was closing down.

"Who knows?" joked Taber, "We probably won't know anything until the others have left through the airlock and we wonder where they've gone!" The four of Them had known for some time that the resignation was afoot, but had not told Us.

"Well, there goes our chance for scientific credibility," I said.

When asked what the SAC's beef was, Iain Prance says, "I did feel that the powers that be were not listening enough to what the scientists were saying, and that was the real problem with the whole thing. They had a Science Advisory Committee, but they weren't listening to, or hearing, what was being said."

Gaie has never understood this, as she had banged day and night on her computer keyboard, churning out reports, press releases, different versions of the research plan, all manner of communications for the SAC members. So focused had she been on feeding the

SAC and media's need for information that the four of Us thought she was shirking her duties in the Agriculture.

But that issue, it turned out, was not the only problem. Years later, Tom Lovejoy would tell me, "I was never convinced that I was being leveled with by John Allen. You'd get one answer one day, and another answer another day. I remember at one point suddenly wondering whether I could ever take John at face value. He was an actor, so I could never tell when he was acting and when he was being real. Literally."

On the day of the final SAC meeting, John approached Tom with tears in his eyes to proclaim his respect and admiration. Tom found himself wondering whether this sudden show of affection was real. He added, "It was so much John Allen's intellectual baby, and he was not the one to think about it scientifically, and that was an inherent problem. It's very hard to tell the parent that they can't have as much as they want to do with the baby."

A press release from Ed Bass' office cited "personality conflicts" as the cause of the mass SAC resignation. It quoted Tom Lovejoy saying, "Although the committee feels the scientific research of Biosphere 2 has made good progress, I and several members of the committee found the working relationship at times to be frustrating. I'd like to reiterate that the whole committee retains its enthusiasm for what Biosphere 2 can contribute to the field of closed system ecology."

The press seized upon the news. "Biosphere or Biostunt?" was the headline in *Time*. The article "reported" that the "seal has been broken several times in the past year and a half," making Biosphere 2 into "just another greenhouse." The article continued, "Now the veneer of credibility, already bruised by allegations of tamper-prone data, secret food caches and smuggled supplies, has cracked. . . . [The] two-year experiment in self-sufficiency is starting to look less like science and more like a $150 million stunt." So pitiful was the reporting that several papers wrote that the project had closed.

Some of the world's leading scientists had given the project a thumbs-down, and that was a calamity; Biosphere 2 was now untouchable.

The SAC gone, NASA blacklisted Biosphere 2. Their official position was that they did not want anything more to do with the tarnished project.

I, of course, blamed John and Margret. I thought it had been a serious weakness on John's part that he could not compromise just a little, to bend just a little to the recommendations, perhaps demands, of the SAC. But Tony Burgess sees it a different way:

"He was just being honest [when he alienated the SAC]. Ultimately he hated them. He was antiestablishment and antiscience. How could he have gotten along with them? Even he could not pull off that theater. Even John could not betray his deeper self that much. Maybe that is not a failure of nerve; he could not pull off the dishonesty. In some ways that is a positive thing. The mistake was having a SAC in the first place, or one that was to give academic credibility to it, because you clearly did not have the people to deal with it, because it got into a power struggle."

On the inside of Biosphere 2, the "all-out war" with the SAC, as Gaie had come to call it, had completely destroyed any vestige of friendly relations between the two of us. She thought I had masterminded the SAC's dissatisfaction by filling their heads with lies, as did Laser, despite the fact I had never spoken to a single member outside organized meetings.

I was beside myself with fury at Mark, for I felt that he of all people, being the chairman of IE and on the board of directors of SBV, could have prevented the SAC's alienation. It was an unfair judgment because few people could have prevented it, given the personalities at play. My accusations of being John's yes-man outraged Mark.

None of us made any allowances for the fact that we were all hungry, exhausted, and felt like hell. I certainly was not at all forgiving of Mark, Gaie, or anyone else inside who I thought was behaving stupidly, even though they were all locked up, all suffering from lowered oxygen, and all reeling from the media and the scientific community savaging the project.

While the SAC's resignation reverberated through the project, Sierra and I were trying to prevent a food import. We looked like skin and bones. Roy wrote in his monthly medical report to Mission Control that if we did not stop losing weight we would either have to reduce our work load or bring food in.

Sierra suggested that we eat less grain and more green bananas until the sweet potato harvest came in. Roy gave the medical thumbs up. We all watched our weight and not the way I do now. I did not tell Roy when I lost another pound. I hoped I would gain it again before the next medical weigh-in.

The first snow fell in huge slushy flakes. The white did not last, but the clouds remained for days and days. The sweet potato fields were at a standstill. They did not seem to be growing at all, so the badly needed harvest was receding into the future.

Nonetheless, some of us were actively preparing for our lives after Biosphere 2. Mark Nelson was taking university courses to prepare for his masters degree in watershed management at the University of Arizona. A professor proctored one of his final exams, watching Mark take the test via video linkup.

Gaie and Laser were starting a nonprofit organization called Planetary Coral Reef Foundation to monitor and improve the health of coral reefs around the globe. The four of Us whined to each other about how their start-up efforts seemed to be impinging on their available crew time—they showed up less and less at the Agriculture.

Roy had been preparing his life after Biosphere from the very first day. "I want to enter a scientist and exit an artist!" he announced to Barbara Smith, with whom he had embarked on a performance art piece, "The Twenty-first Century Odyssey." Roy was readying several exhibits of his work in Los Angeles galleries.

Taber, Linda, and I participated in filming escapades for Roy's music video, including marching in formation across the Savannah in time to his song "Ecological Thing," with crutches as our guns and neck braces placed on our heads for caps.

I did not understand much of his artwork, but it was a welcome distraction and entertaining.

There is a well-known phenomenon that hits people who have had a pinnacle experience, been the focus of attention, and do not have anything to throw their energies into once the experience ends. They fall into a sort of postpartum depression.

Taber and I had heard that some of the early astronauts became alcoholics after their monumental trip around the Earth or to the Moon. Where once they had been making history, they were now doing nothing important, or so they felt. In effect, their life was over, at least until they picked up the pieces and started on a new endeavor.

However, those who had something to go to that they considered useful and challenging fared just fine. *Rapid reentry stress* was the clinical term coined by those who studied people returning from a winter in the Antarctic.

Taber and I discussed this on occasion as the days, weeks, then months passed. What were we going to do when we got out? It had become clear that neither of us wanted to work at Biosphere 2, and we were sure there would not be jobs waiting for us here after we finished our two years. I had received an offer to go to Purdue University, but academic life seemed stale after my adventures to that point.

Instead, the two of us hatched a plan to start an aerospace business. Our dream was to position the company to take part in what we still hoped would come to pass in our useful lifetime: a mission to put a base on the Moon, Mars, or wherever.

We wanted the company to fill some of the gaps we had seen at Biosphere 2. One weakness was the lack of a fully integrated design approach. While the project boasted of being a cross-disciplinary endeavor, in reality the various disciplines had a horrible time communicating, which had led to some of the design flaws we were now living with.

We were convinced that, had the soils scientists, the engineers, the atmospheric chemists, the computer modelers, and the ecologists all communicated better, we would not have put too much compost in the soil and we would not now be trying to figure out where oxygen was disappearing to. We wanted to bring together the skills and strengths of the aerospace industry and biological sciences, and have them work seamlessly together.

We needed to recruit some aerospace engineers, so we turned to the International Space University. As an alumnus, Taber had been posting commentaries about life on the inside on the university's network, ISU-Net. Many students and alumni read and responded to his posts. One of them, Ray Cronice, who worked at that time at NASA Marshall Space Flight Center in Huntsville, frequently communicated with Taber and decided to go on the biospherian diet for three months to show his solidarity. (Needless to say, he lost weight and was thankful when the three months were up.)

These online conversations were a creative outlet for Taber, encouraging him to put Biosphere 2 and his experiences in the context of space exploration. About this time, Taber posted a comment about butchering the pigs that mentioned how tough they were. Some guy called Grant Anderson from Lockheed suggested we give the pigs alcohol: "My grandfather gave their pigs half a bottle of bad whiskey before butchering them to make their meat tender."

"If we had half a bottle of whiskey, even bad whiskey, we would not be giving it to the pigs," wrote Taber, "We'd be drinking it ourselves!"

After a few back-and-forths about how to make booze from the only foodstuff we had in excess, beets (which unfortunately did not result in a recipe), a fast friendship was born. We chatted with

Grant about our plans to start a company where biological sciences and aerospace engineering worked together. He was intrigued.

So we started a business with people we had never met in person. On February 20, Grant, Dave Bearden (another ISU alumnus), and Max Nelson (also ISU), Taber, and I began filling out the paperwork to incorporate Paragon Space Development Corporation. Launching an aerospace firm would be a challenge we could throw ourselves into when our cloistered life in Biosphere 2 was over, and it gave us some needed distractions during siestas and evenings.

Linda had been corresponding with Kevin Kelly of *Wired* Magazine, who invited her and her seven fellow inmates to participate in a party that he was organizing with Stewart Brand of the *Whole Earth Catalog*: Think Globally, Dance Locally.

We were linked by PictureTel, watching hundreds of people in a gallery, chatting, looking at photographs hanging on the walls. Every now and then someone would come over to the camera that fed images to us, and chat. We were clearly wallflowers.

Kevin then gave an upbeat presentation about Biosphere 2, whereupon a woman shrieked a rebuttal, calling the Biosphere rude names, never mind that we were all listening. Others in the crowd finally shouted her down. Meanwhile, outside the gallery doors, other people were handing out anti-Biosphere leaflets. Later we found out they were affiliated with some of the folks who had been fired from Biosphere 2 or otherwise left disgruntled just before Closure.

A couple of days later, all eight of us linked via PictureTel with some friends who were with a group of whale watchers at Dana Point, California. Forget whales—I could not take my eyes off the wide-open ocean in the background. The sun had been shining for the past week. El Niño had apparently left us, and the crisp light made me lust for long walks on the beach.

Taber's and my house was taking shape, and the architect or the builder came to the window every week or so with photos or items we needed to choose. A couple of weeks earlier Mike, the builder, had held up a knotty pine door. This time it was faucets.

We had become so used to the Biosphere 2 system we used instead of toilet paper—which was not allowed because it could not be made inside—that we wanted to include it in our home. It was simply a hand-held water sprayer on the end of a flexible stainless steel hose that we used rather like a bidet. Mike put pictures to the window.

"This one, this one, or this one?" he asked.

We must have been the perfect homeowners—he could limit and direct our choices, we could not get in his way on the job site, and we were finding the whole process entertaining. I have never understood why people say that building a house can be the biggest strain on a relationship. But then again, unlike some couples I have talked with, we just could not get too worked up about having the perfect faucet. Taber and I mulled over the pictures for a couple of minutes, knowing full well that there must really be thousands to choose from.

"It fits stylistically, it's functional, and it's within the budget—go with that one" said Taber, pointing to the one in the middle. We definitely applied the *Apollo*-era motto of "the better is the enemy of the good" to home building. In fact, I try to apply it to my life. It saves so much unnecessary angst.

Finally, some good news blessed the Biosphere. On March 1, Dr. Jack Corliss became the new head of research and development at the project. He was a kindly gentleman, with a shock of graying hair, bushy gray beard and a deep voice. A student of the origins and evolution of life, he had worked with NASA on programming supercomputers to explore the mathematics of evolution. He led the 1977 ALVIN submersible diving expedition to eight thousand feet below sea level, where he discovered living organisms surrounding gaseous vents.

He showed us breathtaking photos of a whole ecosystem of creatures living in complete darkness. It was an ecosystem that was not based on sunlight, like ours in Biosphere 2, but on the heat and chemical energy of hot gases escaping from the earth's mantle through the vents.

It seemed too good to be true—he was someone who understood both the space and ecological sides of the Biosphere.

The only thing tempering my enthusiasm was that Jack, an academic, was to report directly to John. I wondered how much autonomy he would be given, and if he would have what it took to stand up to John. Only time would tell.

On March 12, we all wished Margret a happy birthday over the video and the next day was my turn to celebrate. I opened the two gifts remaining from those that my family had sent in with me. It did not matter what was in them; it was delightful to open a gift.

An angel watched over Taber and me at Biosphere 2 during our enclosure, and her name was Frances Felix. She was our ears, eyes, legs, and hands outside our world. She managed our finances, paid our few bills and spied on our contractor to make sure he was building our house to her satisfaction. She was entirely loyal, and impartial—she also handled Gaie and Laser's affairs. Frances called us her "babies," and she felt she was incubating us for two years until we hatched from our cocoon.

The afternoon of my birthday, Frances and Taber had concocted a surprise. She brought thirty of my friends to the window. They sang "Happy Birthday," waving a big sign. It brought tears to my eyes to have so many people show their support. Inwardly I thumbed my nose at management.

A few days later the four of Us went out to dinner at Le Bistro, our favorite French restaurant in Tucson. Actually, we bought dinner for a couple of friends with the proviso that they take a video phone and show us their meal, mouthful by mouthful.

So they showed us a beautifully set table with a bowl full of mussels, then peppered steak and halibut.

It was almost obscene, a biospherian peep show. Bernd took pictures of huge bites of juicy steak going into his mouth, while Kathy described the meal in exquisite detail. The owner teased us with a view of his most outrageous dessert, chocolate cake layered with mousse, covered in spikes of white chocolate resembling icicles, from which it got its name, Winter Cake. And finally, they all toasted us with champagne.

"Only six months to go," I sighed.

"Exactly six months, two days, and . . ." Taber stopped to look at his watch, "ten hours and eighteen minutes."

"But who's counting?" laughed Linda. There was no question that the countdown to the end of the mission had begun.

Crops were growing again, and finally, finally, we harvested the long-awaited sweet potato field, which alleviated our fears of needing to bring in food. Barring another pestilence like the broadmite attack, it looked like we would make it on the food we had and those promising harvests to come.

Time's dull drumbeat sounded out by chatter on our two-way radios, the bell for meals, and calls to peanut breaks. It lulled us into a stubborn acceptance of our fate, that socially, psychologically, emotionally, even ideologically, we were in a pickle and had to stick it out to the end. None of us dared say anything about it openly lest we unleash the demon that we had to that point kept chained and subdued.

We must have all felt it lurking just beneath the surface, ready to lash out. Sometimes its claws were visible in acerbic comments, in pinched lips, a pallid face, the stiff gait, in dreams of beating others to a pulp.

On occasion I knew the demon was entering the room behind me even without turning to look, the hairs on the back of my neck bristling. And I'd look into its eyes and see the suffering there. That was the truth of it. We were all suffering, hurt beyond words.

Sometimes I sat in my room, wrestling with all the inner voices.

There was the whining simp, "I'm hungry, I'm tired, my back hurts. I don't want to do anything, I just can't go on!" Then there was the angry demon, "I am so fucking mad I just want to scream. Those bastards have fucked us all so badly I don't see any bloody reason why we should stay in here."

The angry demon had a hard time fully expressing itself; no self-respecting Englishwoman fully expresses anger—it is just not done. So then the frozen mummy does not say anything. It simply sits there, a dull lead weight, numbing the mind. Sometimes it was not me weeding the wheat, but the frozen mummy, like a zombie pulling out the unwanted plants. Other times the angry demon was raging inside, while I cleaned the algae scrubbers that purified the Ocean water.

While I was digging in the dirt of my heart, searching for the strength to continue, I sometimes unearthed a jewel and glimpsed the inner peace I had heard of, like the time I was weeding in the Desert. The place no longer resembled a desert at all, but, as we were informed by Tony Burgess, the Desert's designer, it had transformed into a coastal sage scrub, the kind that one sees in areas that have been disturbed by humans—shrubby, overgrown, and straggly. In our attempt to create a desert we created the very thing that ecological purists would consider a blight on the landscape—a disturbed ecosystem.

I crouched in the sand, ripping out large weedy plants. Grasses, sage, and broom had become gargantuan from the constant dripping of water from the roof during the winter, when the plants should have been dry and dormant.

Like an archeologist, I was gradually uncovering the Desert still growing slowly under the burgeoning mass of unwanted foliage. My senses came alive with the pungent fragrance of the Israeli plants Tony had planted for us as natural perfume. I crushed the brittle leaves, gritty between my fingers and thumb. The sound of the pumps and crickets echoed cavernously against the glass and steel, not a melodic sound, but vibrant, vital.

I saw the many shades of greens and browns in the tangled shapes of cacti, yuccas, and shrubs that glowed against the stark white spaceframe. I experienced their beauty and understood their role in our world. Descartes was dead and there was no longer a barrier between me and nature, for I was nature.

At that moment there was no anger. There was no hate, no yearning. Time had no meaning in the normal sense of the word. I was not running from anything, or galloping towards some imagined target. I just was.

And for that moment I was completely at peace.

In that place of peace, I watched my mind take off on flights of fancy. I observed my hands pulling the plants out of the ground, their now-useless roots dangling. I felt empathy for those plants, but then thought that it was only natural—I was participating in the cycle of life, killing some plants to make room for others to thrive. Then I realized the cycle was totally artificial, human made. Gods and goddesses of our own world, we decided who got to live and who died.

I wondered whether Biosphere 2 was an attempt at the ultimate expression of control—the bottling of nature to thoroughly conquer it once and for all—quite an irony for a project that honored Gaia and natural self-organization. For, I mused, if we can cage nature, tame it, and train it to work for us, then we have overcome a distinct glitch in our quest for species immortality.

After all, every living thing's primary innate goal is to have offspring to perpetuate the species. Biospheres simply promise to take that one step further: the species can ensure that it has an environment fit to inhabit.

Living snug in miniature biospheres, *Homo sapiens* could survive a planet that will someday become uninhabitable, whether through our own doing, or through the work of the universe hurling a giant meteor at the earth, or through Sol's eventual death. It may take billions of years for such a crisis, but, by taking the long view,

Homo sapiens would have found species immortality and a way to spread our seed throughout the universe. What more could a living thing strive for?

At that moment, fantasies of biospheres floating through space, keeping cities of humans alive, seemed perfectly plausible, and a lot better than floating in a tin can, as David Bowie crooned about Major Tom in his spacecraft. Whether future space was to be populated with sterile, overcrowded tin cans or biospheres, I only hoped that the people and their cultures would have evolved to get along better than we had. Perhaps, I concluded, we humans are continually striving to conquer nature because we cannot conquer ourselves.

John thinks human potential is realized in part through becoming cooperative agents of the biosphere. However, man versus nature or man cooperating with nature are two sides of the same coin. In classic Cartesian thinking, humans are set apart from nature.

But in Biosphere 2 we were at once separate from it, managing and observing the Biosphere through intelligence, and also part of the very ecosystems that gave us life.

While John diagramed the biosphere and technosphere as two separate but interacting entities inhabiting the same globe, I see them as a continuum, the technosphere embedded within the biosphere, an extension of our nature. When we send Viking to Mars, or Voyager outside the solar system, they are envoys of our biosphere; they are not separate from it, but its sensory organs reaching out into the universe and sending back information about the cosmos as if they were our own eyes on a hugely, impossibly long tether of radio waves.

Are termite mounds not a part of nature? Are the tools that monkeys use to catch the termites not natural? "But that's different, they are all natural materials on a small scale. Humans have made things that are not found in nature," a voice says.

And other living organisms have not? Life has created geology.

Plankton made the 250-meter-high White Cliffs of Dover, for which England was nicknamed Albion (meaning white). They piled up on each other for millions of years.

Our petrochemical industry is built on the backs of dead critters, squashed and heated by geologic forces. That we think we can alter our environment to the point of rendering ourselves all but extinct hardly proves that we are separate from nature.

Our early ancestors, the anaerobic prokaryotes (bacteria that thrive in the absence of oxygen), did this when they oxygenated our atmosphere. No, we have not reached the status of gods and goddesses—we are simply another natural agent altering our environment as life has done for billions of years, and will continue to do, with or without us.

The difference is that we are self-aware. We can alter our own actions. We can be catastrophically dumb, but we can be nobly and creatively intelligent.

We have a choice.

The prokaryotes did not.

Earth Day found several biospherians engaged in giving a virtual tour of our world to over two million schoolchildren, who linked with us via satellite for the event. Sponsored by IBM, this was one of many educational programs that we had all at some time participated in.

Biosphere 2 provided a powerful platform for teaching about the natural world of Biosphere 1. "The earth is too huge for any but the most brilliant minds to grasp," Kathy Dyhr said. "But there we encapsulated it and sixth graders could get it just by looking at it and hearing one or two sentences like, 'Nothing goes in or out except information and energy.' And they would know all the consequences of that. You could not destroy part of it and expect to have a good life. It was the perfect way to make the preciousness of life so apparent."

The oxygen level had not conveniently leveled off at just over 16 percent as predicted by John's report, but had continued to decline. For months, Taber had been working with Jeff Severinghaus, one of Wally Broecker's graduate students at Columbia University, to uncover the cause of the loss. By now the two molecular detectives were hunting for twenty-two metric tons of missing oxygen. They were under considerable pressure because if they could not find such a vast quantity, then Biosphere 2 was undone. It would prove, as some critics proclaimed, too complex to be knowable.

Jeff and Taber calculated the amount of oxygen and CO_2 in the atmosphere inside the sealed structure at Closure, added and subtracted amounts pumped in or leaked out and stored, accounting for the changes in oxygen and carbon dioxide atmospheric levels. Biosphere 2's atmosphere had lost over 8 percentage points of oxygen, and only gained approximately 2 percent of carbon dioxide. The oxygen loss and the carbon dioxide increase should have been equal.

Early on in the investigation, Taber and Jeff came up with two plausible explanations. In both scenarios, soil respiration played a role. As the microbes decomposed the wealth of compost in the soils, they used oxygen and produced carbon dioxide.

In the first scenario, soil respiration accounted for only 17 percent of the oxygen loss, the amount accounted for in the CO_2 scrubber, stored vegetation, and what was in the atmosphere. They postulated that some as-yet unidentified oxygen-eating ogre was consuming the rest.

In the second scenario the soil respiration accounted for close to 100 percent of the oxygen loss. This scenario called for a carbon-dioxide-eating monster that was somehow taking carbon dioxide out of the air. But which was it?

Initially they focused on the direct consumption of oxygen through reactions with nitrogen or sulfur in the soils and water, but analysis showed that there was not enough of either to account for even a miniscule fraction of the loss.

Iron in the soils could be effectively rusting, which would use up oxygen. They also came up with an "oily soil" scenario, whereby natural lipids or other material in the soil would use up more oxygen during the process of decomposition than the amount of CO_2 they produced. In either case, the amount of oxygen loss in the soil would be more than the carbon dioxide produced.

So Taber trudged out to the Savannah and the Agriculture and measured the oxygen in the soils down to five feet. He discovered that the oxygen consumption equaled the carbon dioxide production, so Taber and Jeff crossed those scenarios off the list.

However, Taber and Linda had also been measuring carbon dioxide soil efflux all over the Biosphere since shortly after Closure. Taber and Jeff had ample data to peruse, and they confirmed that the soil was producing large amounts of CO_2. The data did not give them a precise measure of the amount of CO_2 produced, but it certainly pointed to soil as being the culprit. So the two sleuths moved on to solve scenario 2. Where was all that carbon dioxide going?

They could come up with no reasonable solution to the question of what could hold captive tons of carbon dioxide. The only notion they had was that the rain and subsequent droughts in the Desert soils were turning the carbon dioxide into carbonates. It was the only hypothesis they had.

However, the pattern of changing levels of carbon dioxide in the air did not support the hypothesis that the drying and wetting of the soil was linked to removing carbon dioxide from the air. Nope, the carbon dioxide did not seem to be in the Desert soils.

They had reached a dead end.

One morning, Jeff was chatting on the phone with his father, a respiratory physiologist at University of California, San Francisco, who had studied CO_2 and its effects on humans and the environment for decades. Jeff was explaining his frustrations with the inquiry. His father mentioned that concrete was known to absorb carbon dioxide, which was a problem for the construction industry in extreme cases, as the process could weaken the concrete.

There were indeed copious amounts of concrete in Biosphere 2. Perhaps Jeff had hit pay dirt.

So off Taber went to collect concrete samples. He looked like a skinny Rimbaud, armed with a huge drill, pressing his hip against the red handle to push the coring-drill bit into various walls of the Biosphere. Freddy did the same on the outside for comparison.

Even now, when I walk through the Biosphere, I search for those one-inch diameter holes with the extraction date and Taber's initials adjoining them, looking like biospherian petroglyphs.

Lo and behold, upon analysis the carbon dioxide was found sequestered among the concrete molecules, masquerading as calcium carbonate, or limestone. It had reacted with the calcium hydroxide within the concrete.

In order to confirm the finding, Taber and Jeff performed numerous other tests, including isotopic analysis. Carbon comes in several forms, or isotopes, and by measuring the amount of $C13$ in the various components of the Biosphere that interacted with the carbon dioxide, such as the plants and the scrubber, they could determine whether the excess calcium carbonate in the concrete really came from the soil-produced carbon dioxide.

Imagine that the soil is a huge dispenser of yellow and red M&Ms. Inside the dispenser are more yellow M&Ms ($C12$) than red ones ($C13$) because of the ratio of $C13$ to $C12$ found in the soil. The atmosphere catches all the M&Ms as they constantly spew out of the dispenser in the same ratio as is stored inside and carries them throughout the Biosphere to the plants, who prefer yellow M&Ms to red ones.

If there was no one else eating the M&Ms, and because the atmosphere is not bringing very many M&Ms at a time for the plants to eat, they would eventually be forced to eat the red ones, which would otherwise build up in the atmosphere. But, the concrete also loves M&Ms, and has no particular preference for color.

So it snaps up whatever the plants leave. As the red M&Ms are not left to build up in the atmosphere, the plants continue to take more of their favored color, while the concrete continues eating indiscriminately.

If this were what was actually happening inside Biosphere 2, then the concrete should contain more red M&Ms, or C13, than the plants. Furthermore, by knowing how many M&Ms, or carbon dioxide molecules, had floated through the air (which Taber and Jeff knew from their earlier calculations), the guys figured how many red M&Ms should have been in the concrete versus the plants. They took samples of the concrete, the plants, and the CO_2 scrubber and measured the C13 in each sample, from which they could account for all of the red M&Ms that had been dispensed by the soil. And sure enough, they confirmed that the concrete had more C13 than the plants. It had been stealing CO_2 from the atmosphere.

A brilliant piece of scientific detective work had solved the mystery of the missing oxygen, demonstrating that the Biosphere, at least this time, was complex but still knowable. The two could only have solved this riddle in a hermetically sealed environment where every molecule could be accounted for. Serendipity played a role, as so often happens in such detective work—without Jeff's father commenting on the cement we might still be wondering where the oxygen had gone.

What was particularly heartening about the analysis was that it proved that the oxygen loss was not an innate failure of enclosed artificial biospheres. It was a tractable problem and we understood the solution: the concrete could be coated with sealant to prevent the carbon dioxide from diffusing into it.

Laser and the engineers outside the Biosphere were already preparing to do just that at the end of the two years, and the sealant would prove to work well in the second closure.

The excessive carbon dioxide produced by overfeeding the soil microbes with compost could be eradicated by putting less organic matter into the soil. Using calculations similar to those Jeff and Taber used to determine how many oxygen molecules had been lost, it would be possible to calculate how much compost and other organic matter to put in the soil so that the atmosphere would be balanced. For budgetary reasons we would not be able to change out all the soil with a different composition, seal up Biosphere 2 and test it again. So, we couldn't be 100 percent sure with unequivocal certainty that it would work, but there was no reason to think that it would not.

This conclusion raises an unfortunate aspect of the Biosphere 2 project that Lynn Margulis and Dorian Sagan pointed out in their 1987 paper entitled *Gaia and Biospheres*: "[Biosphere 2's] success will have to be defined negatively, as an absence of failure."

We could not be 100 percent sure if Biosphere 2 "worked" until it had been operated flawlessly for many years, and even then, how would we know that some unanticipated problem would not arise from genetic bottlenecks or incomplete nutrient cycles? And what constituted "working"? As Margulis and Sagan wrote years earlier, "Eight biospherians may live inside [Biosphere 2] for two years, but will their continued good health in isolation be a definitive success? What about ten years of closure? Or if they survive inside for a century, producing children and grandchildren?"

Every departure from 100 percent material closure had to be viewed as a failure. In a society where failure is seen as a negative event, and success and failure are viewed in black and white with no gray tones, we were set up for failure. From this perspective, Biosphere 2 could not succeed—something had to go wrong, as we had attempted something entirely new.

In order to succeed, Biosphere 2 had to be seen as an ongoing experiment because in science, an experiment is only considered a failure if nothing has been learned—and we had learned much already. An experiment may disprove a hypothesis, it may result in unanticipated findings, but as long as something is derived from the experiment, it is successful.

This was why so many of us who loved the project with all our hearts had placed so much emphasis on its scientific credibility. This is why we were so furious that it was slipping through our fingers.

After leaving Biosphere 2, Taber and I successfully tested the notion of maintaining the oxygen and carbon dioxide balance with one-gallon-sized sealed ecosystems. Some of them have been going for nearly ten years, completely sealed with small crustaceans living and breeding inside them.

Biospheres work as life support systems. However, until artificial biospheres have undergone extensive testing, I would only bet my life on one that had physical/chemical backup systems.

Along with the oxygen loss, we were still contending with food production issues. Seven of the eight of us were losing weight again, albeit slowly, but it heightened food anxieties. I fumed if another biospherian had put more food on his or her plate than I had managed to cram on mine.

Regina North at NASA, with whom Taber had set up a psychology program prior to Closure that Margret had kyboshed, occasionally chatted with Taber and me. She had spent time performing psychological studies on crews in the Antarctic, and told us that food stealing and hoarding were often problems, even though they do not have food shortages. People squirrel away food in their rooms, apparently to give them a feeling of security.

Well, I did not admit to her that I, too, had stolen food. Linda would put Purina Monkey Chow (one inch-long pellets made mostly of extruded grains) in little bowls for the galagos. One bowl sat, so temptingly, in the Orchard. Most of the time I resisted, but sometimes my hand just found itself in the bowl and I swiped a monkey nut. They were quite sweet.

Another of my dirty little secrets is that on a couple of occasions, Taber and I just had to alleviate our grating hunger. An open bag of a grain called sorghum had mysteriously appeared in his lab supply storage area, a small room up a few stairs in the hallway outside the lab itself. We tiptoed up the stairs and into the room in the dead of night and stole a handful of sorghum. We cooked it up in a beaker on the lab hot plate. On about the third occasion, shortly after our conversation with Regina, we were nearly caught. We had just finished pulverizing the sorghum in the lab's grinder and were standing in the dark, quietly waiting for it to cook.

Suddenly the door opened and Mark walked in. We could hear him, but thankfully he never walked around the little wall that formed an island in the middle of the room, or he would have seen us standing like statues with eyes as big as saucers. We would have had a difficult time explaining why we were there in the dark with sorghum cooking in a beaker.

I wish I could say that we never did it again, but we did, about seven or eight times in all (I did not document my lapses of integrity), until the bag of sorghum disappeared as miraculously as it had appeared.

However, I want it on record, right here and now, that I never ever stole anything from the fridge, anyone else's leftovers, or the captain's birthday cake.

Roy's girlfriend, Yasmine, a beautiful black-haired Berber from Morocco working at the project, was fired on May 21 on what she said were trumped-up charges. She was so upset that she did not arrive at her on-site house to clean it out by 5:00 p.m. as requested. So the local sheriff and a couple of project staff members began the job—an extraordinary violation of private property. When Yasmine finally arrived, the sheriff tried to have her sign a legal document stating that she owed the project large sums of money. The devastated woman disappeared and no one could find her for days. She never came back on site.

Roy was upset. He made a remarkably simple yet moving art piece to express his sorrow, which he showed at our next art festival only a week later: a series of Polaroid shots of Yasmine at the visitors' window in every month of the closure period leading up to that moment. Sometimes there was brilliant blue sky and she was in summer dress. At others she was bundled up surrounded by snow, wearing a stylish French beret. The slide show flashed one month after another with the striking face of the young woman beaming through the window until the last slide, where the window was empty and forlorn.

The day after Yasmine's disappearance Roy made an unsettling accusation to Linda, Taber, and me. Since the beginning of the experiment Roy had encrypted his e-mail so the management could not read it. If Taber or I ever needed to e-mail something that we considered sensitive, we asked Roy to send it for us.

Now he was certain that management was tapping our phones and listening into our conversations. It sounded feasible, but there was no hard proof. Nonetheless, sometimes in jest I would address them in the middle of a conversation with friends on the outside.

If it was true, and someone was indeed eavesdropping on our phone conversations, I considered it a bizarre invasion of privacy. The Biosphere was our home. If it were not true, it was further testimony to the unraveling of our objectivity.

Once our mission ended, I asked Gary Hudman, who spent much of his time in Mission Control, if he thought our phones were tapped. "Oh yeah, no question."

After Yasmine's departure, Roy became understandably sullen, although he still talked with her often by phone. Taber and I took it upon ourselves to cheer him up in the manner that Roy would appreciate—he loved women, not in a lecherous or demeaning way, but in an honest, yet passionate way. And women loved him. So, late one evening we called Roy to the visitors' window.

"Roy, Roy, could you please come down to the window, we have someone we'd like you to meet."

"At this time of night? Who is it? I'm working," he growled.

"Oh, we're sure you'll enjoy meeting this person."

"I'm busy."

"Come on Roy, it won't take long. Just for a minute"

He begrudgingly agreed, and stomped into the little curtained area in front of the window, looking quite irate. Outside stood a gorgeous woman, with a heavy-set man standing about six feet away, looking ready to pound anyone into the ground at a moment's notice.

"Roy, meet Tina and her bouncer. Tina's going to do a little show for you."

Tina turned to a boom box by her feet, turned it on and started to sway and gyrate to the disco music that crackled over the phone. As she started to disrobe, Roy beamed from ear to ear, and his bald head lit up like a beacon. His mood was a good deal better once Tina was done pressing herself to the window and had left Roy with cheery visions of an exotic dancer's beautiful body twisting and pulsating before him.

May had brought sun, sun, and more sun, just the way Arizona should be. The crops were growing and the Wilderness biomes were thriving. The Savannah was about to go dormant for the summer, but at the moment it was in full, neck-high splendor. The transformation from the previous dead, flat-to-the-ground and brown blades of grass, to green blades that were so long they would stick up my nose if I slouched, was boggling.

At times like these I reveled in the experience of being consciously a part of the day-to-day biospheric cycle. As I stood in the grass, I knew I was breathing the oxygen the plants produced and they were absorbing my carbon dioxide (as was the concrete). Our relationship with the other living organisms in our Biosphere was equally symbiotic. While it seemed we held dominion over them by our ability to turn the rain on and off and rip out unwanted plants, our lives depended on them. If the plants were unhappy, as some of them had been in the Agriculture, we suffered. We were in fact on equal footing, we were in this together; we affected and relied on our Biosphere, and it affected and relied on us.

Having experienced such a direct connection to my life support system, and the glass and steel wall around it demonstrating its finiteness, I suppose I was readily able to accept the notion that Biosphere 1 is finite and that humans could be operating on such a scale as to be affecting its functioning. In 1993, global warming was much discussed, but many people still could not believe that we humans would have the arrogance to presume that we could change planet Earth, Gaia.

But our Biosphere was on a human scale. I could walk around it in ten minutes and I could map it out in my head or on a napkin. When events unfolded quickly in our Biosphere, as they did when CO_2 rose more in a single day than the earth's has done in millennia, it spurred us into immediate action. We could not be political about it, we had to put our differences aside—we hiked out to the Savannah and cut down all the dead grasses and fired up the CO_2 scrubber. The intimacy of our Biosphere allowed me to see that we are all biospherians of Biosphere 1, and all creatures are biospherians, too.

Sometimes I am left breathless by our struggle to come to terms with what scientists are beginning to call the Anthropocene era, a new age in our earthly biosphere's history where humans are altering the global environment, the vast atmosphere we all breathe. The enormity, complexity, and, until recently, the plodding pace of change in Biosphere 1 makes it easy to put problems off for another administration.

In Biosphere 1 there is no turning down the thermostat to stop the Siberian tundra from defrosting, with great belches of greenhouse gases likely to spew into the air. In Biosphere 1 no analogous huge concrete wall will miraculously soak up the excess carbon dioxide as it did in Biosphere 2 (although scientists are attempting to construct the equivalent of our scrubber system on a mega-scale).

In Biosphere 1 there is no door to escape through if the going gets too rough, although some claim that space colonization provides one.

In Biosphere 1 we like to blame our leaders for the mess, just as in Biosphere 2 we alternately blamed Mission Control, the management, the SAC, or Ed. In Biosphere 1 we are now living the human experiment.

Gazing through my room window on May 31, I watched deer below eating the precious landscaping again. The night before I had seen an owl on the center arch of the Agriculture at dusk as the first stars were coming out. How I longed to see the stars with no bars between me and the sky! The pull between finishing out our two years and the yearning to be outside kept growing stronger.

Chapter 20

ARE WE THERE YET?

W e had just over three months to go. The end was finally in sight, and it looked like we were going to make the two-year target. But on June 5 we nearly didn't.

At 5:00 p.m. the Biosphere went silent. No pumps churned; no air handlers blew. The computer battery backup systems began beeping. They had lost outside power and were draining their batteries.

Half an hour later the air handlers still lay idle and we turned off almost every computer to save battery power.

A huge fire had swept across the desert just south of the Biosphere, shutting down both the road and the power grid. Our Energy Center had three generators, any one of which should have automatically picked up the load. However, generator 3 was down for repair, and generator 2 died for some reason, leaving only generator 1. It tried, but overloaded the main breaker panel.

Seven engineers, electricians, and Energy Center operators crowded around the panel trying to get it to hold the current. Linda ran to the command room to watch the Global Monitoring System, the computer system that recorded and displayed the temperatures and other environmental parameters throughout the Biosphere.

The temperature was rising. Every few minutes, Linda's voice came over the radio, counting up as the temperature rose in the Rainforest, the highest biome and therefore the hottest.

"Ninety-one degrees." It was a race against time to get the power back on before the temperature became so hot that it would kill or injure many of the plants and animals, forcing us to leave before our two years were up.

Linda continued counting. "Ninety-six degrees."

The engineers, electricians, and Energy Center operators continued to punch buttons, scratch their beards, and discuss frantically how to solve the problem. Every time they tried to bring up power it blew the circuits.

"Ninety-nine degrees."

They decided to bring one section of the Biosphere online at a time, so we ran about turning off all the air handlers, pumps and anything else that drew power.

"One hundred four degrees, guys. How are you doing? It's getting hot in here."

Finally, they flipped the switch outside. One by one, the air handlers whirred into action as we reset them, and the pumps sputtered to life, and the air started cooling.

We all breathed a collective sigh of relief when we heard the noisy equipment stirring again. We knew how lucky we had been. Had the power failure occurred in the middle of the day, instead of the late afternoon, we would all have had to leave the Biosphere after only twenty minutes or so, and our experiment would have ended.

By now we were well out of the "third quarter," the supposed hardest part of any sojourn longer than six months. We were creeping toward the day we would walk out the airlock for our reentry into Biosphere 1.

One major benefit of seeing the end was that we knew with increasing certainty how much food we would have to finish out the two years. That allowed us to increase portions slightly.

We were now eating around 2,200 calories per person per day, and some of us were gaining weight. Our metabolisms had slowed so much from the calorie restriction that with even a small increase in food our bodies responded as if it was spring after a winter's famine, and we had to pack on as much fat as possible before the next winter's dearth. The gains were slight, but they relieved our bodies. Our rising energy helped lighten the mood inside.

Despite the continual hunger of the past two years, our Agriculture had proven much more productive than the most efficient tropical farms, which can support around three people per acre. We could have comfortably supported seven people on our half-acre farm, more than four times as many as the best farms could feed. Photon for photon, our Agriculture was only 6 percent less productive than NASA's highly efficient space systems.

We had also produced a cornucopia of different food, something important to future space colonists, compared to the narrow choices NASA's systems would provide.

Everyone was now so busy with preparations for the end of the experiment and the five-month transition period between our and the subsequent experiment (which we simply called Transition), that morale rose. There was little time for brooding. All the operations manuals, covering everything from how to operate and maintain the waste recycling system to how to set up for a press conference via satellite, had to be updated with everything we had learned. The crew that replaced us would rely on these manuals to know—technically at least—how to live and work within Biosphere 2 during their ten-month mission, scheduled to begin in March 1994.

All the research that we had begun during the past two years had to be completed any loose ends tied up—and documented, ready for publication. Gaie, Mark, and Linda began inventorying all creatures in the Biosphere to ascertain extinction rates.

The study would be completed during Transition and would show less than 20 percent extinction, more than 60 percentage points lower than Howard Odum had bet Linda at the start of the two years. Much of the extinction was loss of insect species that had been overrun by ants and cockroaches, as occurs on small islands.

Crazy ants, *Paratrechina longicornis,* pillaged almost every other insect species, eating their eggs and out-competing them for scarce resources. I would watch in amazement as a band of ants, each less than a millimeter in length, dragged a dead roach or cricket up the vertical steel wall of the basement to somewhere more than ten feet above the ground. On more than one occasion I saw the ants attack a live beetle, overcome it and drag it up the wall.

The termites had also succumbed to the crazy ants. Ironically, the ants ate the silicon glazing sealant, not the termites I had tested. Once I stood at a window in the Desert waving at people outside when I glanced down and saw a trail of black dots disappearing into the silicon and not coming back out again. I suddenly felt like I was standing stark naked in public. What if the visitors saw? I abruptly turned and rushed away, not wanting to draw attention to what I assumed would be a black trail outside.

Mission Control immediately fixed the teeny hole. All in all, the crazy ants created five such leaks during our mission, the Transition, and the second manned mission.

The ants and cockroaches in Biosphere 2 are far less prolific today because the ecosystems have matured. It is akin to building a new aquarium. When it first starts up, the water chemistry is all out of whack, but over the course of a couple of weeks it suddenly comes into a dynamic balance. The aquarium industry has a term for it—the aquarium *pops.*

In Biosphere 2 it has simply taken a little longer, ten years in fact. As I toured there recently with Tony Burgess, he crouched in the Rainforest and pointed excitedly at a skeletonized dead leaf on the ground, a sure sign that something had been eating it. He dug around in the leaf litter and showed me tiny snails, millipedes and sow bugs all munching their way through life, recycling the dead. Finally he looked, his round, fair-skinned freckled face framed by a now graying beard, and exclaimed in his impassioned nasal voice, "It took ten years, but the detritivore community has developed. It just took time."

Back in mid-1993, preparations for Transition moved into high gear. Each one of us systematically reviewed our areas of responsibility and drew up long lists of recommended improvements that Gaie and Sierra submitted to Mission Control.

To help boost crop production, the Transition crew was to hang banks of high-pressure sodium lights over particularly shady places in the Agriculture. Sierra would hunt for a larger threshing machine to cut down on food processing time. We had become so infuriated with the inefficiency of some of our equipment that we had taken to putting dry bean pods and wheat on the ground and stomping on them, crushing them underfoot until the seeds came loose. I felt stupid doing it, and it didn't work very well, either.

People who worked in the greenhouses outside were beginning a beneficial mite-breeding program so that the next crew could import hordes of them to fight off the terrible broad mite that had killed all our white potatoes, a program that would prove very effective.

The Transition team would also put together a system of ropes and pulleys like a sailor's boatswain's chair to haul a researcher into the Rainforest canopy to study CO_2 flux and leaf photosynthesis at different levels within the forest. It would turn out that the plants absorbed more CO_2 near the top of the canopy than towards the bottom, a phenomenon later discovered in natural forests.

Just like getting ready for a visit from mother-in-law, we dashed about our home, cleaning surfaces until they shined. We started with the CO_2 scrubber, which we would not need to use again during our stay.

It brought a sense of the impending end to realize that there were things that we would not be doing again. We started ticking them off. No more CO_2 scrubber crews. The last kids had been born. We began planting crops that we knew we would not harvest; they would be someone else's chore.

A second crew of seven trained on the outside to enter the Biosphere five months after we exited. Norberto Alvarez Romo, the head of Mission Control during our two-year stint, was to be the crew captain. With him would be John Druitt, an Englishman who had worked with IE for many years in the Puerto Rican rainforest; Matt Finn, an American graduate student doing his thesis on marsh ecosystems; Pascal Maslin, an Aussie who had lived and worked on the *Heraclitus* as the chief engineer; Charlotte Godfrey, an American who had been managing the marine systems in the Biosphere's research center outside; Rodrigo Romo (no relation to Norberto, but many thought they were cousins), a Mexican chemical engineer; and Tilak Mahato from Nepal, who had a degree in agriculture and grew up farming much the same way we did in our Agriculture. They made a colorful, young, and energetic team of people.

Aside from the technical aspects of Transition and the next mission, we discussed the group dynamics and psychology of being hermetically sealed with a couple of the new team members. We described in detail the unnerving psychological effects and the splitting of the crew, warning them to prepare themselves for it.

They responded much as we had prior to Closure. "Come on, mano," laughed Rodrigo, "We all know each other. We're friends. We know how to deal with this. It's no big deal."

In fact, they suffered many of the same problems we did. For some reason, it seems impossible to convey the severity of the psychological and social challenges in isolated and confined environments—they have to be experienced to be believed, especially by Type A, self-assured individuals. But unlike us, the second crew had a psychologist on call during their closure.

We still tended to the ninth biospherian's every need. The Agriculture was becoming overgrown with weeds from the warmth and long days of summer. It was all I could do to keep up, even with all-day crews helping pull the pesky grass out of peanut fields. Gaie had recruited Taber several months earlier to help her weed the coral reef, and now she added me to the roster.

Weeding in a wetsuit, breathing long and deeply through a diving regulator, hanging in the cool water over the corals, was a welcome respite from weeding bent over hour after hour in the Agriculture.

Key chemicals in the Ocean were too high and acted like nutrients for algae, which was growing like a red and brown lawn in places, covering the slower growing corals with such deep shade that it risked killing them. So Gaie, Taber, and I became coral gardeners. Happily this got me off the hook for cleaning algae scrubbers, an utterly onerous task.

Our coral reef resembled reefs in populated areas of the world that have been damaged by overfishing and nutrient-laden water from agriculture and cities, making it a poignant tool to study what happens to reefs in such an environment and how they adapt, regrow, or die.

Sierra had seeded and then re-seeded rice when the first batch died. She set up an experiment to discover whether it was the dreaded root fungus, pythium, that had exterminated them. It was.

I felt a sick sense of vindication, since, when Linda had raised this potential problem, John's response had been to make her jump up and down, screaming "pythium, pythium."

We discovered we could reduce the root disease's ability to survive in the paddy soils by drying them out, so the infected areas were left fallow for several months during Transition. If we had been on Mars, we would likely have left most pests and diseases behind, and we would have inoculated a sterile soil with beneficial microbes. Nonetheless, a backup way to definitively destroy pathogens would be a vital part of any farm system.

On June 2, the primary freezer broke down, so we heaved wheelbarrow loads of frozen bananas, pork, and other food to the backup freezer that Sierra had fought for before Closure. It was lucky she had prevailed or we would have been gorging ourselves for the next week before the food spoiled, leaving us short later.

Laser suspected the freezer's freon had escaped into our atmosphere, but it was apparently harmless at those concentrations. The cockroaches had already dispatched both the primary and backup microwave oven, after chewing through the wires. It is an old wives' tale that goats eat anything and everything; however, cockroaches do.

With only fifteen weeks left to complete a long list of tasks, people were becoming hysterical and tunnel-visioned. Sierra and I fought over the water system again, and whose fault it was that hoses leaked or the basement flooded because someone had left a valve open. Mark and Roy screamed about who had grown more food on the balcony. Laser refused to cook after someone ate the soup Laser had wanted to use to baste a piglet.

Submariners who have been cooped up underwater for upwards of six months report that Short Timers' Syndrome sets in a couple of weeks before arriving at port. There is no reason to hold back the angers and dislikes; soon they'll be out of each other's face. Brawls spontaneously break out in the mess and there are cruel pranks.

We exploded into occasional bouts of yelling, but no brawls, and no mean tricks. Still, I felt that the air would ignite, explode into flames with the spontaneous combustion of so much suppressed anger, fanned by the grueling schedule. I had the unnerving feeling that someone would hurt someone else at any time—not that I expected an out-of-control biospherian to come charging out of a closet, knife in hand, ready to stab me or some other victim in the chest.

I was more concerned about the subconscious effect of so much hatred. We were all wary, on our guard, watching our backs. Sometimes I would look up as I walked to make sure no one lurked above me with a heavy object to accidentally drop on my head.

The literature is replete with examples of social problems developing during extended isolated missions. We had apparently done pretty well with our seemingly intolerable situation. In a 1993 paper on psychology in enclosed environments including Biosphere 2, Debra Faktor of George Washington University reports that "onboard an ocean research vessel, $50,000 (U.S.) worth of equipment and two years of data were thrown overboard by the crew following a dispute over whether to use a refrigerator to store scientific specimens or beer. In a 1970 Antarctic expedition, one crew member got into an argument and shot and killed another over wine taken from a trailer."

In Biosphere 2 we managed to avoid damaging each other physically. We despised each other, but we still respected each other's humanity.

Cleaning up the huge mounds of biomass that were heaped all over the basement was simply bizarre.

In a scene straight out of an old sci-fi movie, we stuffed the dry biomass into black nylon bags about eight feet long and three feet in diameter that lay on the lung's white-painted steel floor.

The dust and mold spores hung in the air, forming fog. Everyone wore masks to filter out the particulates while each person made an enormous black plastic-encased worm from our carbon bank. The echoes of our movements and attempts to speak made a cacophonous racket in the white metal room, while the black membrane of the lung hung eerily over our heads.

It was a black-and-white scene. If only our situation in the Biosphere had been so black and white.

We were primping the Biosphere, and she was beginning to gleam in preparation for all the visitors who would tour after our reentry into Biosphere 1. However, the crew looked bedraggled. Mark had not packed enough shoes so he was now working in work boots that were more duct tape than leather. His toes poked out the ends. He went barefoot most of the time, as many of us did. Most of our jeans were worn so threadbare that holes were proliferating as quickly as the ants.

Roy looked like a scarecrow, his pants tied at his waist with string to stop them falling down because his belt was too loose. Laser did not bother tying anything around his waist, and was continually pulling up jeans that hung precariously on his hips, waiting to drop to his knees at any moment. Sagging was not yet in fashion for the white male.

Taber wore bright green suspenders to hold up his trousers. My shirts were either torn or covered in stains, and Linda's looked two sizes too big. Along with our orange complexions from all the sweet potatoes, we must have looked quite a sight—hardly the Hollywood image of space-age heroes!

Most jailbreaks occur within the last six months of a sentence, even a twenty-five-year sentence. The tension between incarceration and freedom becomes so great that some people snap.

Unlike prisoners, the eight of us were free to leave at any time. As we were reminded in memos from Mission Control on several occasions during the last few months of Closure, there were five doors, none of which was locked, any of which we could exit at any moment should we choose to pack it in.

We had to make the decision every day to stay voluntarily enclosed. The temptation to sneak out at night to have a breath of Biosphere 1 air, feel the desert breeze on my face and smell the monsoon rain, became almost too much to suppress. Probably the only things that stopped me from cracking open the door and putting a foot on Biosphere 1 ground just to break the tension, if only for a moment, were the lead rings that recorded if the seals were broken on the airlock doors.

It was not simply that we could not wait to be out of the Biosphere, away from the turmoil, and finished with the last inedible lablab bean meal, but also that so much awaited us, much of it unknown. The anticipation was excruciating.

Taber and I had engaged a lawyer to help us find an agent to organize a speaking tour. He had contacted several agents who expressed interest. One of them called the project's publicist, who was apparently also setting up a speaking tour with the same agency.

Suddenly, Taber and I were defending ourselves from threats of a lawsuit—the publicist claimed we were not allowed to give talks, which was news to us. The letter from Carolita Oliveras, the project's lawyer, was utter baloney—a compost heap, Greg Redlitz, our lawyer, called it. The next day all the biospherians received an e-mail notice from the publicist stating that no one had the right to solicit his or her own engagements—also baloney.

However, she made it clear that Taber and I were not going to be able to use any images of the Biosphere unless we were employed by the project. As it was becoming obvious that continued employment was not an option, we entered negotiations with our captors to obtain a license.

In the midst of this battle, Margret kindly offered Taber and me a small bonus upon completing the two years, as she had promised the other biospherians. She was extraordinarily nice, even inviting me to participate in a play she was directing later in the year. That fly on the wall would have had no idea that her lawyer was lobbing threats at our lawyer.

Later we would end up fighting with her to get our promised bonus.

The monsoons arrived, clouding the Biosphere. Raindrops poured down the glass, and hail pounded the roof. The oxygen level was still sinking, and was now at 17.8 percent. We were beginning to again feel the lack. The CO_2 had risen sharply, to a high of 3,250 parts per million, so, reluctantly, Taber turned on the CO_2 scrubber again. On July 15 we harvested our last rice paddy.

With nine weeks to go we were limping toward the finish line but our excitement was building. The atmosphere vacillated between short-tempered and giddy. I had stopped going to Sunday night dinners, but Roy, who continued to attend so he could record the speeches, reported that Laser said he felt that all eight biospherians could be good friends again. That same evening, Gaie stated that she thought we had become a very good working team.

Both statements were ridiculous. A lot of water would have to go under several bridges before I could forgive some of what had transpired during the past two years. Our communications, organization structure, and decision making were in such disarray that I could not agree with Gaie.

I did have to admit that we should all be proud that we had come so far without killing each other, while maintaining the Biosphere. In any event, their comments did signify a softening of attitude, heralding less-scalding interchanges ahead.

Hopes of having found a worthy replacement for John as the scientific director for Biosphere 2 were dashed during an extraordinarily embarrassing Biogeochemical Cycles conference intended to review and suggest Biosphere 2 science projects. In usual management pathology, the people who knew most about the subject on the biospherian team—Linda and Taber—were not invited except to a short portion where they were not supposed to say a peep. However, so much misinformation appeared, that they had to correct the presentations on work done to date.

Finally, Dr. Corliss presented a plan for a carbon-cycle model that was extremely complicated, and it turned out, Wally Broecker, who was advising on the biogeochemical cycle work in the Biosphere, had previously rejected it. Wally is known for his candor, and after Dr. Corliss defended his plan, Wally stormed out.

Tick-tock. Sometimes I'd look at the clock and think, "Jesus, can't the clock move any faster?" Other times I'd wonder how on earth I was going to get everything done.

Seventy-seven days left to go. Several biospherians bragged about the great deals they were getting on their new cars. Taber's and my house was finished and the builder and architect came to show us the photos. Even Sierra, who had loved her time in Biosphere 2 so much she wanted to be on the second crew, admitted she looked forward to being outside.

The airlock's draw grew even stronger now. I stared at it every time I passed. As the monsoon rain poured down the windows outside, I imagined the sound of the Colorado River toads that emerge from the ground at that time of year, mate in a noisy throng that sounds like a cross between a cow and a startled duck, and then disappear back underground until next year.

The number of interviews was picking up. French TV was making a documentary, Dutch TV came to interview the Belgian, Laser. Newspapers and TV programs from around the U.S. came. They thronged the place like hordes of noisy cockroaches crawling over everything, wanting to know everything.

The oxygen was still dropping and was now down to 16.9 percent. Roy was beginning to feel terrible again. On August 26, with exactly one month to go, oxygen was injected into Biosphere 2 for the second and final time.

Four weeks and counting. Roy had been getting worried about our "escape," as he called it—he was sure that somehow the management would mess up our reentry into Biosphere 1. On August 27, he received a note from three of his medical consultants at the University of Arizona. The memo said we were all to remain on the Biosphere 2 campus for three days after our initial exit from Biosphere 2, for medical and psychological evaluations.

They were concerned about the immunological and psychological stresses of an abrupt reintegration to society, which could "contribute to a depress[ed] endogenous stress response capability." Translated into English, we might collapse from the stress of it all, have a panic attack, or generally freak out.

It was an edict—Roy, the head physician on the project, had not been consulted, although the "medical direction and control staff" (presumably Mission Control) had been in on the decision.

My face flushed with panic when Roy showed the letter to Taber and me.

Taber and I had invited all our friends and families to a huge party at our new home the night of reentry. Roy was planning to bring his friends and family over, too. It seemed that hundreds of people were coming to the event and we had booked every room in the hotels in the little town of Oracle where we were to live.

Margret found out about our party because she had tried to make reservations in the same hotels and could not get a single room because we had taken them all. She demanded we cancel our rooms and give them to her. We did not comply.

Missing our own party was entirely unacceptable to us. For several days Roy had heated debates with the doctors and psychologists. Taber even had a go.

It turned out there was no scientific or medical basis for the demand that we stay on site. In fact, as Roy argued, not allowing us to leave would put more undue stress on us, not less. He fought with Norberto and then Margret. But it seemed we would be stuck inside Biosphere 2 while everyone else ransacked our new house without us.

Even with all the excitement, tensions between Us and Them still simmered as evidenced in the destruction of the Butt-wheeled Wagon. Sierra came into morning meeting announcing that it had to be removed from the Plaza. The four of Us refused to have anything to do with that. So Roy set up a camera in the Plaza trained on the wall. Gaie scrubbed the body paint from the white wall, while Laser ranted into the camera about terrible fascist art.

To make matters worse between us, Laser's brand new fire-engine-red Volvo turned up outside one day. Taber and I were convinced that Margret had bought it for him, just as Gaie and Laser were certain that Ed had paid the two of us off to turn Us against Them. Neither was true. But we all filled our information vacuum with assumptions about each other's actions and motivations.

During the final week, we all checked the multiple-choice answers in a Minnesota Multiphasic Personality Inventory questionnaire. The test was originally developed to determine serious pathologies, such as schizophrenia, in mental patients. It has since been used to ascertain profile "signatures" of different types of people such as astronauts, professionals, and prisoners. Dr. Michael Berren of the Arizona Center for Clinical Management scored our tests and concluded none of us suffered mental pathologies.

However, where normally the profiles of men and women are very different, the profiles in our case were almost indistinguishable. Moreover, when correlated with profiles of astronauts and people who winter over in the Antarctic, we all showed close correlations, particularly the women. The test suggested that a certain type of person signs up for adventures like ours.

According to Dr. Robert Bechtel, who also participated in the study, we exhibited the profile of people who do well in the Antarctic and in space. In a paper published in 1997 entitled *Environmental Psychology and Biosphere 2,* Bechtel concluded, "Selection for isolated environments follows an adventurer profile. . . . These scores define people who are somewhat guarded about themselves, highly active, creative, optimistic, sure of themselves, and extroverted."

One of the most important aspects of a personality that is able to deal with the stresses and strains of an isolated situation is a low D score, meaning a low proclivity for depression. Our score was lower than astronauts who had been measured, which filled me with a certain degree of satisfaction mixed with disbelief, considering how lousy I had felt for the past year.

After reviewing the data from the two crews Dr. Bechtel concluded that, "These data do suggest that the Biospherians are selecting or being selected for certain traits that are classically useful in stressful situations."

Our training in the outback, on the *Heraclitus,* on stage, and at the project, however unorthodox, had weeded out those who did not handle stress well, and had narrowed the field to those with essentially the same profiles as the members of the Astronaut Corps. Of that I was proud.

Prior to Closure, Taber had secured a fully funded psychology monitoring and testing program to be paid for and performed by NASA. When Margret found out, she was so angry that Taber almost lost his place on the crew. She said she quashed the research because it would be a complete invasion of biospherian privacy. However, her fury suggested it would also be an invasion of managerial privacy.

Whatever the reason, her decision caused the project to lose the opportunity to learn some things of direct relevance to NASA's spaceflight program. Our crew of eight people was the first and only small group of people to have spent two whole years in isolation voluntarily, and our record remains unbroken.

Dr. Berren's test and exit interview also confirmed something the eight of us already knew: that we were all bulldog determined to see the two years out.

On September 19, Norberto sent a memo to all the biospherians, copied to company lawyer Carolita Oliveros, with whom Taber and I had already had some unpleasant interchanges over our public-speaking engagements. He insisted that we stay on campus for the first three days after reentry for medical reasons.

"So that's how it'll be. You're really going to proceed with this rubbish," I thought. "I'll sneak off site if I have to, run away with my knapsack on my back if they force me to. I am going to be at my own party, damn it!"

Finally, after much arm-twisting, Roy managed to get the physicians to concede that if we signed a release, we would be free to party down.

It seemed we might escape the circus after all.

Reentry minus four days, twenty-three hours, and ten minutes. Having polished the Biosphere until it shined, we now turned to our own appearances.

I had been the crew's hairdresser during the two years. Before we came in, a professional hairdresser had given me a template for each person's hair. Gaie's was simple: comb it straight down and cut a straight line along her back. Mark's was relatively easy, or so I thought: comb a small section of hair straight, hold it in between my index and middle fingers of my left hand, pull it taut so the edge of my fingers is about an inch away from his head and snip.

However, as the two years had worn on, the haircuts got further and further away from the original template, and some biospherians had only intermittently allowed me to touch their hair. So now, I cut everyone's hair (except Roy who did not have any) in front of a television with a hairdresser on the outside demonstrating what I should do. It worked, or close enough.

Friends and family began to arrive. We could see the tents and stage being erected on the lawn outside for the long string of events planned through the day and into the evening. TV vans lined up on the artificial mountain behind the Biosphere. Excitement electrified the air. We had almost done it.

Eighteen hours and counting. All our friends from the International Space University had come from all over the country for the Big Day. They came to an Agriculture window to say hello to Taber and me. We laughed and chatted over two-way radios.

Then came a moment when those people over the age of twenty-five were immediately separated from those under. As a tall man with white hair walked agilely behind the gathered crowd Taber remarked that Timothy Leary had just walked by.

Those above twenty-five rubbernecked to look, those below in unison said, "Who?" Timothy, a friend of Roy's, came to our party the following night—I'm still hearing about it!

My parents came to the window, along with my brother Malcolm, to let us know that they had furnished our house with a bed so we did not have to sleep on the floor. My dear mother was so happy to see that Taber and I looked perfectly healthy. I could not wait to be out and hug and kiss everyone.

People often ask me what I missed when I was inside Biosphere 2. The answer is friends, family, and the intimacy of hugging and talking without the glass in between us. I was not a huggy person before Biosphere 2, but now I am.

After a fitful last night in my room, surrounded by packed boxes of my belongings, we rose early, fed the goats, watered the plants, ate our last porridge, and got ready to walk out. My stomach was full of butterflies as I donned the blue biospherian suit that I last wore the day we entered. It, like everyone else's, had been taken in, as otherwise it would have hung off me like a sack.

Everyone sat, one by one, in the swivel chair in the middle of the command room while I put on their makeup, as I did before every TV interview. Taber stood in the lab window looking down at the hordes of people.

Ten minutes. We walked down the spiral staircase to take up our positions in the mechanical shop next to the airlock. The butterflies had multiplied and were flying up my throat.

Eight o'clock. Two years exactly.

But no call came for us to reenter Biosphere 1. We waited.

And waited.

Finally the speeches were over and there was the call crackling on the radio. Two years and twenty minutes after we first walked into the Biosphere and closed the metal doors behind us, we walked triumphantly out again.

We had made it!

Time magazine wrote in their "Best Science of 1993," where Biosphere 2 was voted seventh in their top ten list, "They may have been the butt of countless Leno and Letterman jokes, but the eight men and women of Biosphere 2 did what they said they would do: they spent two years locked inside the world's largest terrarium and managed . . . to make a significant contribution to ecological science."

Chapter 21

REENTRY

A fter two years living enclosed in the separate world of Biosphere 2, I longed for the warm embrace of friends and family.

But the smells of Biosphere 1 hit us first. Heady perfume, shampoo, soap, and deodorant greeted us before my friends' and family's excited smiles, hugs, and laughter. We had no perfumes of any kind inside Biosphere 2 as they would have interfered with Taber's research on atmospheric components and might have been mistaken for aromatics produced inside the Biosphere. So our noses were hypersensitive to the smells.

The sickening stench of the belching golf cart we boarded to go to our medical check was almost overwhelming. But who cared? We were riding on a vehicle that moved without human power. I was about to eat whatever I wanted to eat, go wherever I wanted to go, and hug and kiss whomever I wanted to, without the glass and steel skin of our world intervening.

Shepelev, the Russian who had sealed himself in the closet-sized room with algae that was the first baby-step towards Biosphere 2 so many years ago, sent a message to Mark, "We congratulate you and your fellow Biospherians on the brilliant completion of the experiment, on your unprecedented heroic deed, which is really a step forward along the way of historic progress."

Yes, we had done it. We had somehow made it through the first two-year manned mission, emerging forever changed by the experience, but thrilled and excited by the incredible accomplishment. Sure, we had had our fair share of problems. But we had met a basic criterion for success—we had proven that a man-made biosphere can successfully sustain life, including human life, for an extended period of time without inexplicably crashing, or devolving rapidly into green slime.

The two primary technical problems of our mission, reduced oxygen levels, and inadequate food production, were indeed serious, but were understood, and attributable to design flaws that could be fixed in subsequent experiments. The third problem— eight people cooped up for two years—was one that even today is an issue for crews on the International Space Station and in the Antarctic. But none of these problems negated the promise of biospheres as life support systems or ecological tools.

I floated through the morning, transported in my own little bubble, all the events unfolding around me. During our medical exam, Taber had noticed the emblem on one of the doctors standing by, denoting that she was with the airlift crew. Because of concern that the stress of reentering the highly contaminated, sensory-overloading world of Biosphere 1 might cause one of the eight of us to collapse, a helicopter stood by to rush us to the hospital if necessary. Taber sidled over and inquired as to how sick he had to be to get a ride on the helicopter.

Before I could say Jack and the Beanstalk both of us were strapped onboard, and looking down from a couple of thousand feet onto the structure that had been our only home for the past two years.

It looked so small. Miniscule.

What a couple of hours earlier had seemed like an enormous area for eight people to care for, with its seven ecological zones, its twenty-eight hundred species, its pumps, fans, and sensors all crammed into just over three acres, shrank absurdly before my eyes. I found it hard to believe that we had stood being inside such a tiny place for so long, especially when compared to the vastness of the surrounding desert.

The day was like a fairy tale. The people who had built our house surprised us by hiring a white stretch limo. It took us to lunch at the new home that we had never seen. Ray Cronice, one of Taber's friends with whom he had corresponded from inside, had brought homemade beer, which Taber had promised would be the first thing to cross his lips. We drank it at lunch.

The first thing I ate was a large strawberry, which I savored—much to my own surprise, as I really do not care for strawberries. A local supermarket presented each of the biospherians with a large basket full of items we had requested at their behest—they were somewhat embarrassed that almost every basket was filled entirely with booze, cheese, and chocolate!

Linda was more imaginative and had asked for gardenias—her favorite flower—and *The New York Times.*

Other friends from the Society for Creative Anachronism put on a show for us, dressing up in medieval clothing and bashing each other over their armored heads with nasty-looking primitive weapons. They came bearing gifts of mead.

Afternoon found us in a bizarre scene with the corporate lawyer, ensconced in her oak-clad office with the single box of stuff that each biospherian was allowed to take out that day. (Apparently the rest needed to be inventoried before it left the Biosphere, never mind that it had not been inventoried on its way in.) She would not give any of us our belongings, and was rude to the point that my normally unflappable, demure mother jammed her foot in the door and forced her way in to demand my belongings. The door slammed open to reveal shabby biospherian knickers strewn around the desks, apparently being counted.

My father, who had run his own business for many years, kept saying, "What a strange way to go about personal relations." Taber and my lawyer brother Malcolm called our lawyer. I was furious, but also could not help thinking the whole episode hysterically funny.

It got even weirder.

Taber, who had been given a tape recorder by a documentarian friend to record the day for posterity, ran into Norberto. Norberto became close to apoplectic when he patted Taber on the back and discovered the strange lump. He was certain that Taber was trying to catch something on tape so he could later blackmail him. Norberto confiscated the tape. Taber called his lawyer. The lawyer said, "Do nothing." But, when Taber threatened to take his complaints to the media throng, his tape promptly reappeared.

Roy escaped the demeaning box debacle. He had arranged for his family to snag his box and whisk it away as it was taken out of the Biosphere in the late morning. In fact, he had concocted the plan with his lawyer, so sure was he that management would attempt to take his valuables, particularly his video and audiotapes documenting the two years.

He had instructed his daughter Lisa to hide the box under her bed in the Biosphere Center's hotel room and to crash through the front gate if anyone tried to stop her as she left for the airport. Lisa, not knowing that indeed some of us underwent searches of our possessions, said later, "He was so paranoid. He was disturbed when he got out—deranged even."

In the early evening, we belly-laughed at a skit that Jay Leno had taped for us. It made up for the fun he had had at our expense during the mission (which we had secretly found highly entertaining). While we were suffering from the effects of low oxygen, he was joking on his show about how we were like those insects we collected as kids and kept in little jars. "Only," he said of us, "Didn't they know you've got to poke holes in the top of the jar to let air in?" This evening it all seemed a quite normal Jay Leno shtick until the end when he was exiting toward the camera. He ran into a piece of clear Plexiglas and stuck there like a bug on a windscreen. Tears poured down our faces. We loved it.

Then it was finally time to relax, and spend time with friends and family. How we partied! At one point Roy and Taber spontaneously toasted in unison, "To freedom and dignity!"

But one person hung in the back of the crowd and would not come and talk to us directly. Dave Bearden was one of the people with whom we had started our aerospace business. We had talked at great length over the telephone, corresponded effectively via e-mail, and occasionally seen images of each other. When I asked Dave what was wrong, why he would not come and talk with Taber and me, he said that it felt peculiar to be in the same room with us. "I keep wanting to run next door and call you on the phone!"

My experience, on the other hand, was that for the first time in two years, a cold pane of glass did not block the warm touch of a friend's hand.

I had escaped the Biosphere.

Chapter 22

THIRTY-SEVEN BRANDS
OF KETCHUP

I was reborn into Biosphere 1.

I wondered if I would remember how to drive. I slipped behind the steering wheel of my parents' rental car, imagining kangaroo-hopping down the road like an embarrassed teenager. But no, I gracefully took off, speeding down the highway with the landscape sliding by, the wind buffeting my face. I was racing across vast tracts of space. I was no longer confined. I was free!

I marveled at the thirty-seven brands of tomato ketchup on the supermarket shelves, and danced eagerly from foot to foot trying to decide which one to buy. I gawked at the ocean of produce. I wanted to throw myself into the bakery cabinets and roll around in the cakes and truffles like a pig in heaven.

And the walls of wine were simply intimidating. "Taber," I gasped, "Just look at all that fermented grape juice. Where do we start? We should have a bath in the stuff." The world outside Biosphere 2 was a sensory orgy.

Shopping took hours. I read every label on every package. Still, I wondered where everything came from and what was in it. After two years of everything being recycled I filled an entire rubbish bin full of packaging just to put food in my refrigerator.

I could taste the fat in foods and all desserts were far too sweet for my cleansed palate. I engulfed bags of green crisp apples.

I even enjoyed pawing through junk mail. I approached each new activity with childlike innocence and wonder.

But I was constantly late. After two years of walking down the hall for a meeting I found myself dashing across town only to realize that I had not left enough time. After two years of nothing but a two-way radio and pruning shears on my belt, I could not keep track of things. I lost my sunglasses, my handbag, my car keys. All the sensory input overwhelmed me. Even the exhaust from my car's tailpipe was nauseating, reminding me I was back in the world where technology leads, unlike Biosphere 2, where technology was subservient to life, and nothing was allowed to produce noxious waste. I was glad to be living in the country where I could sit and watch the cactus grow and listen to the grasshoppers to clear my mind.

I suffered alarming crying fits for no discernible reason. Once, I was driving so slowly through the small town of Oracle, tears pouring down my face, that a policeman pulled me over. When he saw my red face, he turned around, walked back to his car without a word and drove away.

On another occasion, Taber and I headed out to a sunset lookout to have a romantic moment. I sat in the car and cried for forty-five minutes straight, while Taber sat patiently waiting for the storm to blow over. We missed the sunset.

I had never been a crier. I could count the number of times I remembered crying on one hand—until now. The fits came without warning, triggered by I knew not what.

But the pain and anguish poured out of me in heaving gasps, until I was drained and exhausted and the tears dried up. I never sought help from a professional and slowly, over the next few years the crying fits would recede from once a week, to a monthly collapse. Eventually I would have my "crises," as I jokingly came to refer to them, infrequently.

Finally, ten years after reentering Biosphere 1, I spoke with

Susan Schwartz about it. She agreed with my amateur assessment that it had been some form of post-traumatic stress disorder.

In retrospect, we should all have had a regimen of medical and psychological consultation, for not only had we all undergone extreme psychological pressures, physically most of us were in poor shape. I felt fragile, something I had never been my entire life.

Chapter 23

APRIL FOOLS

B ut the story was not over. On April 1, 1994, I sat at my makeshift desk in my new adobe home overlooking a vast desert valley and Biosphere 2. The phone rang.

"Hey, Jane. Guess what? John, Margret, and the whole gang have been booted off the project." Steven "Chubasco" Storm, the caller, had been a longtime member of John's group until about a year before. He had been in charge of the tissue-culture laboratory at the project.

"Ha, ha, very funny, good April fools, Chubasco!"

Early that morning, Chris Bannon had sat at his desk in his Beverly Hills office and took a call from his brother, Steve, who was in Fort Worth. SBV had hired Bannon and Co., their boutique investment bank, to get the Biosphere's budget under control. The project had lost a purported twenty-five million dollars in 1992—more than most of us imagine earning in a lifetime.

Steve first met with the project financial officers in April 1993. Despite efforts to cut budgets, many millions were still pouring out of the place. In the October following our exit from Biosphere 2, Steve and other financial advisors had flown out to meet with John and Margret to try and come to an agreement as to how to fix the problem. Something had to be done.

But John and Margret thought Biosphere 2 was theirs to manage. Margret politely but firmly told the men they had to leave the property. Apparently they needed the men's rooms, as a big conference was about to start. Steve felt they had been banished. In documents filed with the U.S. District Court in Fort Worth, Bannon said that he had been forced to suspend an audit of the project's finances that December, after Margret denied auditors access to the property and ordered employees not to speak to him.

In February 1994, a small group including Steve and John flew to New York and John presented his case to several of the large investment houses. None of them took the bait. John and Margret refused to allow other management to come in to get the Biosphere on track financially.

Meanwhile, Ed's people were compiling a record of alleged financial mismanagement, that included, according to the court documents and newspaper accounts, Margret diverting funds for her personal use, including having SBV pay to remodel her house, thefts from the six gift shops, kickbacks, and apparent sweetheart deals with subcontractors.

Ed finally decided to act. Steve flew to Fort Worth to help Ed's team hatch a restructuring plan. With what was approaching a quarter of a billion dollars at stake, and emotions running so high, it had to be done ... with an element of surprise.

"Grab a suit and tie, bring a secretary and one of the accountants, rent a Lincoln Continental and meet me at the Tucson airport." That was it. That was all the information Steve gave Chris.

Chris sat outside Tucson International Airport wondering what his brother was up to. It was the days before stringent airport security and Chris's Lincoln sat with three others, lined up along the curb. Finally, Steve walked into the chilly spring morning with an entourage of about ten people, mostly men in dark suits—lawyers and accountants from Fort Worth—and Terrell Lamb, Ed's publicist. Martin Bowen was with them, a cigar-toting short Texan with a big attitude, who was one of Ed's business advisors. He was president of a foundation dedicated to building a performing-arts center in

Fort Worth, and the local business community had high regard for his management skills.

"Head downtown and stop at the courthouse," Steve ordered. The caravan of Lincolns pulled out of the sleepy airport, and filled up with more men in dark suits at the U.S. Courthouse. An SUV with federal marshals wearing flack jackets and packing guns joined the caravan. They looked as if they were headed for a drug bust.

They headed north out of town. An hour later, the cavalcade pulled up to the guardhouse at Biosphere 2's main entrance. A marshal leaped out of one car to prevent the on-duty guard from radioing ahead and announcing their arrival. Chris stayed behind with the guard.

In the lobby of Mission Control, three secretaries sat at their desks. Five men in black suits charged in with the marshals. "Where's Gary Hudman?" one suit demanded, flashing an official-looking piece of paper.

All three women blanched, wondering what could possibly have happened to require armed guards on the scene. One of the women showed them to the computer room. "He is in there," she whispered.

Gary sat where he always sat at this time of day, at the Global Monitoring System computer screen. He was checking on the status of each biome in the small world that lay just across the narrow service road from where he slumped, squinting through his glasses at the tiny numbers on the screen. Someone began pounding on the locked computer-room door. (Biosphere 2 information was considered highly proprietary.)

Gary leaped up and opened the door. When he saw suits and guns he immediately pushed on the door to close it. "You can't come in here!" he shouted, leaning all his meager weight against the door and gaining a perverse thrill from the ridiculous spectacle.

The suits smashed open the door, throwing Gary back into the room.

Ten years later Gary recalled the scene. "They handed me the temporary restraining order. I didn't believe them. I thought it was some kind of joke. Everyone was gone. John, Margret, Gaie, Laser. I thought because they were all gone somebody was playing a trick on them. I guess it was a trick."

Apparently, Steve, Martin, and the lawyers in Fort Worth had decided that they had to take control of Gary first, because he was running the controls of the Biosphere. Two guys with guns were posted to watch his every move. Gary laughed to himself. He knew what they didn't—that he would never have intentionally damaged the Biosphere.

Steve headed upstairs in Mission Control with more suits and guns. They took control of Margret's and the other offices, not allowing anyone to so much as touch a keyboard or a power switch. Several people closely affiliated with John and Margret were escorted off-site like convicts and told not to return.

It turned out that Ed Bass had taken control of Biosphere 2 in no uncertain terms. It had been a perfect strike with John, Margret, Gaie, and Laser all out of town at once.

Chris Helms, the head of the project's public relations department, was sitting in his office. As he did every morning, Chris had called Ed's publicist, Terrell Lamb, when he arrived. Her husband answered the phone. "Hi Chris, how's it going?" he said nonchalantly.

"Oh, fine. Is Terrell there?"

"No, she wants to talk with you but she can't come to the phone right now. How's everything there?"

"Oh, everybody's gone, it's a slow news day."

"Oh, okay. I'll have her call."

Chris hung up. Within minutes he heard Terrell's voice in the hallway and he stood up, confused.

Terrell stopped in his doorway and said, "We're here to take over."

Chris grabbed her, fell into her arms, and started crying uncontrollably with relief. For over six months he had been telling Terrell that he wanted to leave the project as it was impossible for him to do his job. For over six months, Terrell had been begging him to stay, telling him, just as she had been telling Taber, Linda, Roy, and me, that, "Something's going to happen, just wait. Things will change."

She had kept her promise.

News of the takeover flashed to the town of Oracle, and reports came of a sign erected by the side of the highway that read, "Ding! Dong! The witch is dead."

Five days later, the crew was tucked up in bed in the Biosphere, having started the second mission just over a month earlier. The day they entered on March 6, 1994, Stephen J. Gould, the renowned scientist and prolific author of popular science books, said, "Who remembers the second man on the moon? I think it was Buzz Aldrin. The first time is a symbol; it's an act of brave pioneers. But I think it takes a different and special kind of courage to be second. So I salute all of you who are going in today, because you are converting what is a grand idea into a simple daily utility which is what it needs to become. The eyes of the world aren't on you anymore, but the eyes of science are, and the hopes of the world are."

The five men and two women were to stay inside for ten months, testing new systems and continuing the science we had begun, as well as watching over many new projects.

It was just after 3:00 a.m. on April 5. A female voice crackled over the two-way radio, "Norberto, Norberto." The voice told the biospherians they needed to leave the building for their own safety as there was no one at the project who knew how to run the facility.

Rodrigo Romo started awake in his Biosphere 2 bed and thought, "Holy crap, that's Gaie? That doesn't make any sense." He then heard the sound of footsteps running up and down the hallway. Finally a head popped around his bedroom door and told him there had been a break-in. All the airlocks were standing wide open, and a glass pressure-relief panel in one of the lungs had been broken.

All seven biospherians gathered at the main airlock, looking out through the two doorways, their metal doors hanging open. Several people had gathered on the other side of the airlock. The two groups could have reached out and touched each other, but they didn't. Nobody wanted to cross what felt like sacred space.

Steve Bannon, whom Martin had appointed interim CEO, was in the group on the outside. He asked the biospherians, "What do you guys want to do? Do you want to come out?"

The response was unanimous: "No."

So they closed the airlock doors, and the biospherians went around the Biosphere closing all the other doors. The doors had remained open for fifteen minutes. Approximately 15 percent of the atmosphere was exchanged with outside air.

Charlotte Godfrey recalls her shock: "I remember standing at the Rainforest airlock which had never been open because it's not a real airlock. It's just a door. And the air was flowing into the Biosphere. I stood thinking it was so surreal. This was not the way it was supposed to be. If you have ever walked into a house that has been robbed, it feels violated. Someone had violated our space."

That someone was Gaie and Laser. They had flown back from Japan, and crossed the desert at night to break open the very structure they had helped build.

The next day all seven biospherians met in the dining room for breakfast. Norberto, who was Margret's husband by this time, as well as a strong supporter of the project, looked like his whole world had collapsed. He was pale, his eyes baggy, and his permanent limp was more pronounced than usual.

Matt Finn walked in clutching a bottle of Jack Daniels, put the bottle on the table and mumbled, "Laser must have left it inside." Norberto looked up and croaked, "Oh, what the hell, pour me." At nine o'clock in the morning the biospherians were doing Jack Daniels shots.

The Sheriff's Department had been called and around fifteen police officers arrived at the airlock. One biospherian thought they should search inside the building to make sure Gaie and Laser were not hiding inside somewhere.

Chris Bannon and the biospherians had to lead a pack of paranoid cops through every inch of the Biosphere. They found no one.

Then lawsuits sprouted. A cockfight broke out in the press and a catfight in the courts. Margret and Ed sued each other.

Larry Hecker, the lawyer for Margret, et al., called the allegations of mismanagement and improper use of funds "grossly exaggerated, overstated and, in some cases completely inaccurate." He told the *Arizona Daily Star,* "The one thing those pleadings do is completely ignore the years and years of slave-like dedication and the tremendous sweat equity that went into this project."

Even Chris Helms was dragged into the fight after Gaie and Laser sued him for an allegedly libelous press release he wrote about the break-in.

Most of the parties settled quickly. But Gaie and Laser faced criminal charges for vandalizing the project. It took five years of depositions and witnesses, but finally the criminal charges were dropped as part of their settlement with Ed.

But the question hanging heavy in the air was, why all this cloak and dagger stuff? Why creep across the desert at the dead of night and secretly break in? To this day, Gaie and Laser say it was a symbolic act. It was a way of allowing the crew to end a mission—should they wish to—that Gaie, Laser, John, et al. had begun but had no way of ending, they felt, other than under the cover of darkness. In 2000, John wrote that Gaie had done "her thankless required duty" to notify the crew of the takeover.

Rodrigo Romo was not convinced, saying to me later, "Next time someone thinks I need saving, please do it at a decent hour, not at three o'clock in the goddamn morning!"

Chapter 24

AND NOW WHAT?

B iosphere 2 almost died in the cradle because of the tumult from the end of 1989 to mid-1994. But Ed Bass, who had owned 50 percent of the project, gained custody, and, thanks to his tenacity and undying faith in the viability of Biosphere 2 as a badly needed tool for global ecology, it survived. Dr. Michael Crow, who at the time was vice provost of Columbia University, saw this vision and, along with the then-president of Columbia, Dr. George Rupp, launched Biosphere 2 into its next career as a model for global-change research.

For eight years, from January 1, 1996 to December 23, 2003, Columbia ran Biosphere 2, naming it the Apparatus, or B2L (Biosphere 2 Laboratory). For those of us who had been deeply involved in its creation or who had lived inside for any extended period, calling Biosphere 2 a mere instrument was derogatory. For us, Biosphere 2 was an entity—part living, part machine—that had taken on a life of its own.

We old-timers cringed as Columbia transformed the Biosphere. They severed the Habitat and Agriculture from the Wilderness. The animal pens below my bedroom, where Sheena had given birth to triplets, became a carpeted exhibit of an energy-efficient Volvo, and the remainder of the Habitat, a museum, with viewing windows in bedroom walls so tourists could glimpse what life had been like on the inside. Our diverse farm grew a single-species cottonwood trees.

In the Wilderness, plastic curtains separated each biome to allow for tighter temperature and atmospheric control. Gaping doorways and fans pierced the skin, so the air that we had considered sacred could be flushed out to enable tight CO_2 control. Eventually the Wilderness was riddled with pathways so that visitors could walk through its innards and imagine what had been—for everyone wanted to know about the people who had lived inside.

One day, a friend still working at the project showed me a poster of Biosphere 2, originally with inset photographs of biospherians at work and smiling faces of the management. The poster he showed me looked exactly like the original, but with un-peopled inset photographs. We had been airbrushed out of Biosphere 2's history.

Nonetheless, dedicated researchers from Russia, Germany, the United Kingdom, and the U.S., who worked under Columbia's aegis, did groundbreaking science in the revamped apparatus. Scientists raised the CO_2 level in the Biosphere's Ocean, thereby lowering the seawater's pH. After measuring coral and algae growth, they concluded that the earth's rising CO_2 levels could severely damage coral reefs around the world. Marine scientists have seen coral growth decrease with elevated CO_2 in natural reefs. This was the first experimental data that not only proved it, but quantified it.

"In a sense, [this discovery] makes the entire project worthwhile," says Tom Lovejoy. "It is going to feed right into how the whole world will end up looking under climate change. As the oceans sour, the vast majority of the organisms that use calcium carbonate will not be able to do so, or they will dissolve. The whole notion that we could make the seas sour has got to give pause. This is not like acid lakes where you add lime. You can't add lime to the oceans; they're too big. . . . It will fundamentally change the oceans."

This change threatens Americans', and the world's, stomachs. New England could lose its clams.

The Rainforest produced another significant finding. Scientists had been concerned that droughts make tropical forests put more CO_2 into the air than they take out. Typically, tropical forests account for as much as 50 percent of the CO_2 taken out of the earth's atmosphere by terrestrial ecosystems. With an increased incidence of droughts that scientists anticipate as a result of global climate change, this would be bad. However, after subjecting the Biosphere's Rainforest to three years of intermittent droughts, the B2L scientists concluded that the Rainforest remained a sponge for CO_2, not a source.

Even the cottonwood trees that we old-timers derided rendered surprising results. Researchers have known for some time that increased amounts of atmospheric CO_2 act as a plant fertilizer, thereby increasing the amount of CO_2 the plant absorbs in order to grow its trunk, branches, leaves, and roots. This, it was thought, would help counteract our globe's CO_2 rise because the world's plants would be growing faster. What the B2L scientists unexpectedly discovered is that the same increase in CO_2 also stimulates the production of CO_2 from the soil. This is not good news, as it would offset some or all of the fertilization effect of the CO_2.

These are but three examples of the discoveries made in B2L, the Apparatus.

In 2005, Barry Osmond, who was the president and executive director of Biosphere 2 from 2001 to 2003, and one of the greatest plant experimentalists alive today, published a paper in the book *Air Pollution and Global Change.* He wrote, "It is clear . . . that the Biosphere 2 Laboratory delivered handsomely as a controlled environment facility for experimental ecosystem and global climate change research. I don't think there is any doubt in Biosphere 2's delivering science that is unique."

The reasons that Osmond, Crow, Bass, Allen, and others support Biosphere 2 for ecological research are control, speed, and scale. Biological systems are affected by many parameters—temperature, humidity, wind speed, light, and soil moisture, just for starters. It is difficult to control most of these in the field. Biosphere 2, however, controls most of the variables, so experiment data are easier to interpret.

Also, things generally happen faster in Biosphere 2 than they do in Biosphere 1. For example, the atmosphere is so small that the CO_2 swings as much or more in a single day than the CO_2 in the earth's atmosphere does in thousands of years. Experiments that would take years or even centuries to perform in Biosphere 1 take less time in a sealed environment such as Biosphere 2.

As for scale, a lot of laboratory-scale research has produced data under highly controlled conditions. But it is impossible to extrapolate from a few cubic meters to the earth's biosphere. Scientists involved in large-system research, such as climate change or geochemical cycling, are beginning to wish they had large-scale, tightly controlled, complex ecosystems to fill the gap.

Walter Adey, the Smithsonian marine scientist who had been in charge of the Biosphere 2 Ocean design, has probably put it most succinctly: "I figured we needed to know how to run these ecosystems in a box. Our whole world is in a box now. Learning to do this was the most important thing that could have been done by science in the tail end of the twentieth century." It is still crucial in the twenty-first.

But Columbia pulled out. Many had scorned the first regime under John Allen and Margret Augustine. But all was not rosy under Columbia's reign, either. Chris Bannon, who remained head of operations the entire time Columbia University ran Biosphere 2, said, "Working for the people at Columbia is as good as it gets. At every level of the organization, people stopped what they were doing and bent over backwards to help me. But here on the Biosphere 2 campus it was much more difficult. I was always going back to Mike Crow and telling him that it was madness out here. It was kind of like the biospherian days. This one was fighting itself the whole time, too."

The details of the endless dramas are unimportant here; suffice it to say that, as in the first regime, egos ran amok and intrigue was rampant. For those of us watching from afar, the similarities between the two regimes were striking, all the more ironic as the two groups held opposing views on almost everything except the value of Biosphere 2–after all, the first group was antiestablishment, Columbia University is the establishment.

Many people, including Chris Bannon, point to what they consider the high operating cost of Biosphere 2 as the primary reason why it is difficult to make a go of Biosphere 2, why it loses money, and why the Department of Energy and the Department of Agriculture refused to provide major grants to the project. It would mean taking big chunks out of other program budgets, stealing from other projects' rice bowls.

Barry Osmond doesn't buy that argument. He, among others, is convinced that large-scale systems for experimental ecological research are sorely needed and horrendously underfunded. In an e-mail to me he wrote, "The running costs of B2L were a small fraction of the agency budgets. . . . I think we had enough experience to state them accurately as . . . about 20 percent of the Kitt Peak [Observatory] administrative grant or 8 percent of one NASA planetary biology outreach project."

In 2002, Mike Crow left Columbia to become president of Arizona State University. "Jeffrey Sachs took over Columbia's Earth Institute, which Crow started, [and under whose aegis Biosphere 2 was run]," Wally Broecker tells me. "Sachs said, 'To run a place that far away is madness' . . . so Columbia shut it down." Just like the first regime, the second one left in a flurry of lawsuits. This time it was because Columbia had broken its contract.

Now, Biosphere 2 still crouches in the searing desert sun. No one conducts research there, the student housing is empty and a few curious people tour inside the biomes, all of which are still functioning, except the Agriculture, which lies fallow. Biosphere 2 is up for sale. Chris Bannon said in June 2005, "The owner [Ed] and us are not ready to give up on it yet. The economy is right. Hopefully there is a major university interested with the right person in control, and I think there is, who can take this and give it a third life. This is the last chance. If we don't make this one work, it's over."

It gives me great hope that, despite dreadful management, and despite, or perhaps because of, the gigantic egos, we accomplished great things at Biosphere 2. Thanks in large measure to Columbia, large-scale closed and semiclosed complex ecological systems are beginning to be seen as credible tools for experimental ecology, biochemistry, geochemistry, and other sciences that encompass the entire globe. And although many said it could not be done, our team conceived of and built Biosphere 2, learned to operate it, and demonstrated that man-made biospheres are possible.

I think, "Look how close we came the first time."

EPILOGUE

The day I walked out of Biosphere 2 was my last day at the project. Taber followed a couple of weeks later. After leaving enough time to make sure that our love would thrive in peacetime, Taber and I were married on the lawn outside Biosphere 2 in the summer of 1994. Our friends on the crew of the second mission looked on from inside the Habitat.

Taber and I went on to build our aerospace company, Paragon Space Development Corporation, with Grant and Dave, who were nuts enough to start the firm with us while we were in the Bubble.

It has not been all easy sailing. As has been our *modus operandi* throughout our adult lives, we learned by doing. We knew so little about what it took to run a company when we first started, that we did not even know what cash flow was, not to mention a profit and loss statement.

But, now we are doing fun things like designing life support systems for spacecraft and equipment for Navy divers. The first animals to go through multiple generations in the zero gravity of space did so in the tiny beer-can-sized biospheres we designed and flew on the International Space Station.

Today we are working to make many identical enclosed ecosystems that all behave the same so that we can address one of the primary problems of Biosphere 2—that there is no other biosphere identical to it with which to compare it.

Just as our beloved Biosphere 2 attempted to address both space and the environment, so, too, do we. My new endeavor, Yogi and Company, seeks to further Ed's notion of ecopreneurship in the areas of sustainable development and global climate change. Unlike many, I do not see the space industry as counter to the environment, but rather, in support of it.

We race motorcycles for fun.

Linda received her PhD in systems ecology from the University of Florida in Gainesville, where she studied with Howard Odum. It turned out that Mark went to the same university at the same time and they sometimes bumped into each other on campus. It was awkward, as the four of Us and the four of Them were not on speaking terms when we parted at reentry. Linda is now teaching at Central Arizona College north of Tucson.

Roy died in 2004. Lou Gehrig's disease finally took him. What a terrible way to go! And what a dreadful irony for a man whose life's work had been to push back death's boundaries, increasing our useful lifespan. Roy's muscles atrophied until his one-time Olympic gymnast's body was puny and shriveled. Finally, he could no longer do the thing I take for granted every moment: breathe.

His resilience was Herculean. The disease had started as the funny tottering gait he had during our two-year closure together. We all laughed then. Before the disease stripped him of all ability to walk—and less hearty souls would have already been in a wheelchair—he went with Safari to Ouagadougou, Burkina Faso, where Africa's largest film festival is held. No one in his right mind would have tottered off to a rough and ready getaway in his state, but that was Roy.

He worked on his scientific papers and his art projects, and he joked, right up to a few hours before he died . . . the first time. The medics brought him back, but it was too late. He was in a coma, halfway between here and there.

Roy had wanted to accomplish what Timothy Leary had not—give the finger to death itself. He arranged to be frozen the moment he died so he could be resuscitated years later when a cure for ALS had been found. But UCLA's top neurologist said that his brain would awaken in a very damaged state, if it awoke at all.

Taber and I joined ten or so of Roy's other close friends and family in a circle around his bed in Los Angeles. Tibetan lamas had sent a sacred cloth to put over his body, which along with their chanting after the moment of his death would, they affirmed, help him on his way in peace. Roy's daughter Lisa was to call them in Prescott, Arizona, once Roy's pulse stopped so they could begin to sing him on his way.

The nurse came in and quietly turned off the beeping monitors and the ventilator, and unhooked the IV. Roy continued breathing in shallow whispers. Lisa stood at the head of the bed and held his skull in her hands so lovingly while she felt his pulse and chanted. We all chanted. Peter, Roy's son, read a Jewish prayer. Lisa chanted.

Peter rang a Tibetan bell to talk with Roy's spirit. We held hands and chanted. It was beautiful, all these people come to be with Roy in the final moment of his extraordinary, courageous, inspiring life.

Roy would have loved the ceremony, with all its theatricality. He would have loved that we all held hands around him while he died. It was so intimate, like our life together in the Biosphere. We were laid wide open for each other to see. It was intimate even between people who despised each other. Being present at a loved one's death is more intimate than making love. Being with Roy at the moment his body surrendered and took its last breath, when his urine drained and his heart gave the final weak contraction, was the most intimate act of all.

It was then that I decided to dedicate this book to Roy, for it was Roy more than anyone who brought grace, humor, and dignity to our existence while trapped in our debacle at Biosphere 2.

Most of the second crew of seven stayed inside Biosphere 2 after the takeover on April 1, 1994, for five more months while the project was restructuring. They must have felt like Cosmonaut Sergei Krikalev, who had been stuck on the Mir Space Station, not knowing when or how he would get back to earth, while the U.S.S.R. disintegrated. After several months in no-man's land, Krikalev finally left Mir and landed in the new nation of Kazakhstan. The biospherians had their feet on *terra firma,* but the mission objective and duration were in the air.

Norberto Alvarez Romo, the crew's captain and Margret's husband, left the Biosphere a few days after John, Margret, et al. were ousted. Bernd Zabel finally got to live inside Biosphere 2, initially only for three days. But he remained inside as the crew's captain for the rest of the second mission. After a couple more months, Matt Smith, a young, energetic biologist who had been working with Walter Adey at the Smithsonian, took Matt Finn's place on the team. Finn had completed his marsh research—the reason he had wanted to live inside Biosphere 2—and now wanted out. The other five crewmates soldiered on.

The second mission lasted six months instead of the planned ten. After proving that the farm could produce enough food for seven people, and after figuring out that the CO_2 problem would take approximately ten years to resolve itself, they exited. September 17, 1994, was the final day that humans inhabited Biosphere 2.

In August 2005, I finally broke more than ten years of silence with the four of Them and the management. I met with Gaie, Laser, Mark, Sierra, John, and Freddy at Synergia Ranch in Santa Fe. Margret was not available to join our conversations.

It was like old times, before Biosphere 2 made us all behave in ways previously unknown to any of us. We laughed about the silly things we had said and done. We patted ourselves on the back for having made it through the hard times.

John gave a simple explanation for what went wrong. "Biospherics total-systems approaches were targeted for scientific extinction by certain 'mainstream' leaders and by Ed and some other corporate powers who consider this science to be inherently dangerous to their superprofits," presumably because when viewed through the lens of closed-systems thinking, the oil industry, for instance, is plainly shown to be ecologically untenable.

"From mid-1991 to April 1, 1994 were the only miserable years of my life," John said to me.

What I will always find perplexing is that John, a Harvard MBA, seemed to ignore the golden rule—the person who has the gold rules. John could not win his fight with Ed, particularly once the scientific community abandoned the project during our enclosure.

But perhaps that's the point. Perhaps he did not want to. Perhaps he wanted to end the only miserable period of his life.

John wept as he told me how he had always loved me and how heartbroken he had been when I had stopped talking with him. "But I thought you had stopped speaking to me," I retorted.

At our meeting in Santa Fe, Gaie handed me a letter that she, as associate director of research at SBV, wrote to Tom Lovejoy and copied to the other members of the SAC six months before they disbanded.

After reading it, I was astonished that Tom did not resign then and there. The letter, a response to the committee's report published in 1992, was a six-page angry diatribe. Gaie called Tom naïve, and intimated that he was sexist, closed-minded, and ignorant.

I felt like immediately e-mailing Tom to apologize, even though it had happened thirteen years ago and I did not personally write it. This was the man who originated the concept of debt-for-nature swaps, and started the pioneering project in Brazil to study the minimum size a piece of rainforest had to be and still maintain a high degree of biodiversity (a term he invented, incidentally).

Reading the letter made me sick to my stomach. It brought the whole tragedy of Biosphere 2 rushing back at me. Biosphere 2 was intended to bridge the gulf between holistic and reductionist science, a union so desperately needed if we are to grasp and solve the global problems of today. The two approaches are coming together, but it would have occurred so much faster had Biosphere 2 shone.

Gaie and Laser now spend a good deal of time on the *Heraclitus*, working on various coral-reef research and conservation projects under the banner "Planetary Coral Reef Foundation," the nonprofit they started while inside Biosphere 2. Sierra is its financial officer. She also works most of the year on IE's project in the Puerto Rican rainforest, growing and harvesting valuable trees in a sustainable and environmentally enhancing way.

Mark received his PhD from the University of Florida in environmental engineering sciences, and like Linda, studied with Howard Odum. He designs and installs biological waste treatment systems around the world.

Margret finally got her architect's license and she now practices in Santa Fe, New Mexico.

And John? Never one to shirk outrageous ideas, he is writing about how humans should be considered the sixth kingdom of life, and a novel set, as he said, "in a place and time where I can say what I want without getting into trouble," fifty thousand years from now in a galaxy far, far away.

Gaie, Laser, Mark, Sierra, and other members of IE started a nonprofit to continue studying closed systems, and built an approximately forty-cubic-meter sealed chamber where they conduct plant experiments, building up to what they hope will be the Mars on Earth project—a prototype Mars base for four people. They spend time at Synergia Ranch, Australia, and the other projects, all of which are now comfortable and welcoming. Things have mellowed quite a bit, but they still have group dinners on Thursday and Sunday nights. The Synergia lifestyle lives on.

Safari settled in the Tucson area, and is still crazy about critters. She promotes the use of rare domestic-animal breeds so we don't lose their gene pool, serving on the board of several national and international rare-breeds associations. She raises a menagerie of unusual types of chickens, turkeys, rheas, pigs, horses, dogs, cats, and any other out-of-the-ordinary domestic animals that come her way.

I lost track of Norberto, though. He is no longer with Margret, and lives in Mexico somewhere.

Gary Hudman stayed on at Biosphere 2 for several years. He now designs software for environmental projects, and is starting a renewable-energy endeavor of his own.

Ed is still a self-proclaimed ecopreneur and philecologist (a philanthropist who works to increase the well being of ecology and the environment). He donates a great deal of time and money to many environmental endeavors, including time in the field on projects for the World Wildlife Fund, of which he is now vice chairman. The Yale Institute for Biospheric Studies that he played a large role in starting in 1990 is still thriving, and he remains an advisor to the program.

He is still unassuming, and still wears cowboy boots and blue jeans, perfectly pressed with fold lines right down the center of the pant legs.

The quarter-of-a-billion-dollar question, for that is about what the whole venture has cost Ed, is, "Was it worth it?" This question has haunted me since the day I walked out of Biosphere 2 in that prickly blue suit. Did we accomplish enough to warrant so much effort, so much money, so many people's sweat and tears, so much excitement and attention?

If I had been forced to answer that question only a couple of years ago I might have said, "Of course," because I had to, because no one wants to think one has wasted years on something of no consequence.

I would have been lying. I was certain we had screwed the pooch, we had done what John constantly warned us all of—we had snatched defeat from the jaws of victory. That was then. Thankfully, time has a way of changing many things.

The story of Biosphere 2 is big and complicated, and it is important to separate the people from Biosphere 2, and from the legacy of closed system science. They are inextricably linked, but separate nonetheless.

Take the people, for starters. For certain, our band of mavericks accomplished the impossible in building Biosphere 2, and then blew it, and blew it big time. Perhaps for the first time we really had something to lose, which has a funny way of making people lose their heads. But we got the thing built and demonstrated its viability. For that we can all be proud.

We should be particularly proud because there is unanimous agreement that Biosphere 2 itself is a marvel. One could argue *ad infinitum* about whether its unique attributes as a research tool outweigh its intrinsic problems. It's not perfect. That's a fact. But Biosphere 2 simply would not exist, nor anything like it (for nothing like it exists even today), had Ed Bass not put up his own mountain of greenbacks to make it happen.

So then I have to ask myself, so what? Did building Biosphere 2 accomplish anything of real significance? Has it furthered the legacy of closed-system science, and is closed-system science itself relevant? It has taken almost fifteen years to answer that question. I am certain Clair Folsom and Vladimir Vernadsky would grin if they could see what has been accomplished here. Biosphere 2 has unquestionably furthered the legacy that began in the early sixties with Clair's handfuls of beach sand in sealed jars, and the Russian, Shepelev, who enclosed himself with vats of algae.

We've put closed- and semi-closed-system science on the map for ecological research and space colonization.

It will not be called biospherics, nor will it be called closed-system science. Nonetheless, large-scale, tightly controlled, complex ecological systems surely will be used for global experimental science. They are needed too badly, and the threat of climate change is too critical, for them to disappear now. Laboratory-scale systems are already used, and other large-scale systems are appearing, such as the $180 million enclosed rainforest in Coralville, Iowa.

Today's heirs say they are doing everything better than Biosphere 2, and I should hope so—we were the first, and trailblazers always make huge mistakes. That they learned from what we did, that it inspired and in some way directs what they are now doing, secures Biosphere 2's place in the legacy of closed-system science, or whatever it becomes known as, and demonstrates its relevancy.

For those involved, the experience seems to have been personally gratifying. Everyone I spoke with who participated during even the most tumultuous and harrowing times at Biosphere 2 made heartfelt statements about how the project changed them and their lives.

John Miller said to me, "I can't think of anything in my life that has been more exciting, except maybe some of my kids when they were born, being right there when they popped out. It was like a birth—a pregnancy, a growth, and a birth when you guys went in there."

Now, more than twenty years after I first went to SunSpace Ranch, soon to be home to Biosphere 2, and listened in awe as astronaut Rusty Schweickart described his first experience of seeing planet Earth hanging in the deep black of space, I can honestly say, "Hell, yes, it was worth it."

Seeing Biosphere 2 as either a success or a failure is fallacious. That attitude oversimplifies the world we live in, and diverts our attention from what is most important—that to dream crazy dreams and work like hell to realize them takes heroism. Chris Bannon said, "Your parents always tell you that you can be anything you want to be, and if you put your mind to it you can do anything you want to do. Don't let anything stop you. To conceive this machine and what it was supposed to do, and to get it done was a miracle. Even with all the madness associated with this place, it got done. It's built, and it's still sitting there. When you walk through that place and see how much thought went into it—all the money in the world doesn't get that thing built. It's a monumental testament to the human spirit. The Pyramids are a testament to the human spirit. The Great Wall of China is a testament to the human spirit. Whether you like everyone or not, those personalities got it done, along with a very honorable guy back in Texas who put his money where his mouth is, and never got a penny out of it."

And if I had to do it over, I'd do it in a heartbeat.

BIBLIOGRAPHY

Note: The bibliography lists only journal articles and books. The many quoted newspaper and magazine articles, along with the personal and television interviews and correspondences, are not shown.

Alling, Abigail, and Mark Nelson. 1993. *Life Under Glass: The Inside Story of Biosphere 2*. Oracle, AZ: The Biosphere Press.

Auger, R., Facktor, D. 1993. *Policy Issues in Space Analogues*. Presentation at the IDEEA Two Conference.

Bechtel, Robert B., Taber MacCallum, and Jane Poynter. 1997. *Environmental Psychology and Biosphere 2. Handbook of Japan-United States Environmental Behavior Research: Toward a Transactional Approach*. Eds. Seymour Wapner, Jack Demick, Takiji Yamamoto, and Takashi Takahashi. New York, NY: Plenum Press.

Bennis, Warren, and Patricia Ward Biederman. 1997. *Organizing Genius: The Secrets of Creative Collaboration*. Reading, MA: Addison-Wesley.

Beyers, Robert J. and Howard T. Odum. 1993. *Ecological Microcosms*. New York, NY: Springer-Verlag.

Bion, Wilfred. 1959. *Experiences in Groups*. New York, NY: Basic Books.

Botkin, D.B., B. Maguire, B. Moore III, H.J. Morowitz, and L.B. Slobodkin. 1979. A foundation for ecological theory. In *Biological and Mathematical Aspects in Population Dynamics,* R. de Bernardi, ed. Proceedings of the Palanza Symposium, *Memorie dell' Istituto Italiano di Idrobiologia*. Suppl. 37.

Burrough, B. 1998. *Dragonfly: NASA and the Crisis Aboard Mir.* New York, NY: HarperCollins.

Burroughs, William S., and Brion Gysin. 1978. *The Third Mind.* New York, NY: Seaver Books.

Dudley-Rowley, M., S.Whitney, S. Bishop, B. Caldwell, P. Nolan. 2001. "Crew Size, Composition, and Time: Implications for Habitat and Workplace Design in Extreme Environments." ICES, paper # 2001-01-2139.

Farson, Richard. 1996. *Management of the Absurd.* New York, NY: Simon & Schuster.

Frye, Robert J. 2002. "Serendipity, Science, and Biosphere 2: A Response to Walford." *BioScience,* Vol. 52. No. 5.

Harte, J. 2002. "Towards a Synthesis of the Newtonian and Darwinian World Views." *Physics Today,* Vol. 55, 29–37.

Kanas, N. 2005. "Interpersonal Issues in Space: Shuttle-Mir and Beyond." *Aviation, Space, and Environmental Medicine,* Vol. 76, No. 6.

Kramer, Joel, and Diana Alstad. 1993. *The Guru Papers: Masks of Authoritarian Power.* Berkeley, CA: North Atlantic Books.

Lapo, Andrey. 1987. *Traces of Bygone Biospheres.* London, England: Synergetic Press, Inc.

Lebedev, V. 1988. *Diary of a Cosmonaut: 211 Days in Space.* College Station, TX: Translation Gloss Company.

Lovelock, James. 1979. *Gaia.* Oxford, England: Oxford University Press.

MacCallum, T.K., and J. Poynter. 1995. "Factors Affecting Human Performance In The Isolated Confined Environment of Biosphere 2." Conference Proceedings, Third Annual Mid-Atlantic Human Factors Conference.

MacCallum, T. K., G. A. Anderson, J. E. Poynter, Y. Ishikawa, K. Kobayashi, H. Mizutani, Y. Kawasaki, J. Koike, K. Ijiri, M. Yamashita, K. Sugiura and L. S. Leigh. 2000. "The ABS (Autonomous Biological System): Spaceflight Results from a Bioregenerative Closed Life Support System." ICES-164.

Marino, B. D. V., and H. T. Odum, eds. 1999. Special Issue: Biosphere 2, Research Past & Present. *Ecological Engineering,* Vol. 13, 1–374.

Nelson, M.T., S. Silverstone, J. Poynter. 1993. "Biosphere 2 Agriculture: Test bed for intensive, sustainable, non-polluting farming systems." *Outlook on Agriculture.*

Nelson, M., W. Dempster, N. Alvarez-Romo, T. MacCallum. 1994. "Atmospheric

Dynamics and Bioregenerative Technologies in a Soil-based Ecological Life Support System: Initial Results from Biosphere 2." *Advances in Space Research.* Vol. 14, No. 11. 417–426.

Odum, Eugene P. 1971. *Fundamentals of Ecology,* 3rd ed. Philadelphia, PA: Saunders College Publishing.

Osmond, Barry, Gennady Ananyev, Joseph Berry, Chris Langdon, Abigniew Kolber, Gunghuilin, Russell Monson, Caroline Nichol, Uwe Rascher, Uli Schurr, Stan Smith, and Dan Yakir. 2004. "Changing the way we think about global change research: scaling up in experimental ecosystem science." *Global Change Biology.* Vol. 10: 393–407.

Osmond, Barry. 2005. "Experimental ecosystem and climate change research in controlled environments: Lessons from the Biosphere 2 Laboratory 1996–2003." From *Air Pollution and Global Change,* Omasa K., ed. Tokyo: Springer.

Orr, James C., Victoria J. Fabry, Olivier Aumont, Laurent Bopp, Scott C. Doney, Richard A. Feely, Anand Gnanadesikan, Nicolas Gruber, Akio Ishida, Fortunat Joos, Robert M. Key, Keith Lindsay, Ernst Maier-Reimer, Richard Matear, Patrick Monfray, Anne Mouchet, Raymond G. Najjar, Gian-Kasper Plattner, Keith B. Rodgers, Christopher L. Sabine, Jorge L. Sarmiento, Reiner Schlitzer, Richard D. Slater, Ian J. Totterdell, Marie-France Weirig, Yasuhiro Yamanaka, and Andrew Yool. 2005. "Anthropogenic ocean acidification over the twenty-first century and its impact on calcifying organisms." *Nature.* Vol. 437: 29.

Paglia, Donald E., Roy L. Walford. 2005. "Atypical Hematological Response to Combined Calorie Restriction and Chronic Hypoxia in Biosphere 2 Crew: A Possible Link to Latent Features of Hibernation Capacity." *Habitation.* Vol. 10: 79–85.

Pierce, C. M. 1991. Theoretical approaches to adaptation to Antarctica and space. In *From Antarctica to Outer Space: Life in Isolation and Confinement,* A. A. Harrison, Y. A. Clearwater, C. P. McKay, eds. New York: Springer-Verlag.

Poynter, J. and D. Bearden 1997. "Biosphere 2: A Closed Bioregenerative Life Support System, An Analog for Long Duration Space Mission, Plant Production in Closed Ecosystems." Kluwer Academic Publishers, 263–277.

Roszak, Theodore, Mary E. Gomes, and Allen D. Kanner, eds. 1995. *Ecopsychology.* Berkeley, CA; Sierra Club Books.

Sagan, Dorion, and Lynn Margulis. 1987. *Gaia and Biospheres*. Boston, MA: Boston University Press.

———— 1989. *Biospheres from Earth to Space*. Hillside, NJ: Enslow Publishers, Inc.

Severinghaus, J. P., W. Broeker, W. Dempster, T. MacCallum, M. Wahlen. 1994. "Oxygen Loss in Biosphere 2." *EOS, Transactions, American Geophysical Union*. Vol. 75, No. 3.

Silverstone, S. E., M. Nelson. 1996. "Food Production and Nutrition in Biosphere 2: Results from the First Mission September 1991 to September 1993." *Advances in Space Research*. Vol. 18, No. 4.

Sipes, W. E., Vander Ark, S. T. 2005. "Operational behavioral health and performance resources for International Space Station crews and families." *Aviation, Space, and Environmental Medicine*. Vol. 76, No. 6.

Suedefeld, P. 2005. "Invulnerability, coping, salutogenesis, integration: four phases of space psychology." *Aviation, Space, and Environmental Medicine*. Vol. 76, No. 6.

Vernadsky, Vladimir. 1986. *The Biosphere*. London, England: Synergetic Press, Inc.

Veysey, Laurence. 1973. *The Communal Experience: Anarchist & Mystical Communities in Twentieth-Century America*. Chicago, IL: University of Chicago Press.

Walford, R. L., S. Harris, M. Gunion. 1992. "The calorically restricted low-fat nutrient-dense diet in Biosphere 2 significantly lowers blood glucose, total leukocyte count, cholesterol, and blood pressure in humans." Proceedings of the National Academy of Science, USA. Vol. 89: 11533–11537.

Walford, R. L., R. Bechtel, T. MacCallum, D. E. Paglia, and L. J. Weber. 1996. "Biospheric Medicine as Viewed from the Two-Year First Closure of Biosphere 2." *Aviation, Space, and Environmental Medicine*. Vol. 67, No. 7.

Walford, R., D Mock, T. MacCallum, J. Laseter. 1999. "Physiologic changes in humans subjected to severe, selective calorie restriction for two years in biosphere 2: health, aging and toxicological perspectives." *Toxicological Sciences*. Vol. 52, 61–65.

Walford, Roy L. 2002. "Biosphere 2 as Voyage of Discovery: The Serendipity from Inside." *BioScience*. Vol. 52, No. 3.

Walford, Roy, Dennis Mock, Roy Verdery, Taber MacCallum. 2002. "Calorie Restriction in Biosphere 2, Alterations in Physiologic, Hematologic, Hormonal, and Biochemical Parameters in Humans Restricted for a 2-Year Period." *Journals of Gerontology Series A: Biological Sciences and Medical Sciences.* Vol. 57, No. 6.

Weyer, C., R. L. Walford, I. T. Harper, M. Milner, T. MacCallum, P. A. Tataranni, and E. Ravussin. 2000. "Energy Metabolism After 2 Years of Energy Restriction: The Biosphere 2 Experiment." *American Journal Of Clinical Nutrition.* No.72: 946–953.

Wilson, Edward O. 1994. *Naturalist.* New York, NY: Warner Books.

ACKNOWLEDGMENTS

There are many people I must thank, without whom this book would not exist. All those who worked way beyond the call of duty to design and build Biosphere 2—I cannot name you all but you know who you are. Ed Bass for your dedication and belief in the impossible, and John Allen for your vision and drive—without both of you, artificial biospheres would still be science fiction. My husband, Taber MacCallum—without your love, support, and gentle insistence this book would still be an idea. Jim Cantrell and Bob Datilla—your goading inspired me to tell this story. Jake Page—your invaluable guidance enabled my book. Everyone I interviewed—thanks for taking the time to tell me your tale, and for encouraging me with the words, "I'm glad someone's finally writing the book." Rebecca Boren, for your help at all hours. Joe Regal and John Oakes, for believing in me. My editor, Anita Diggs, for driving me to make difficult decisions. Chris Bannon and Terrell Lamb, for your patience. Lisa Walford and the Roy L. Walford Living Trust, for throwing open Roy's precious archives.

PHOTO PERMISSIONS

The following photos are used with permission from Lisa Walford, the Roy L. Walford Living Trust: First page (top and bottom), second page (bottom), third page (top), fourth page (top and bottom), fifth page (top), and sixth page (top).

The following photos are used with permission from Terrell Lamb: second page (top), third page (bottom), sixth page (bottom), seventh page (full), and eighth page (top).

The following photos are used with permission from Peter Menzel: fifth page (bottom) and eighth page (bottom).

INDEX